DNA-RNA Research for Health and Happiness.

Jose Morales Dorta, Ph. D

This book was originally titled, <u>The Survival of Coding genes inside the jungle</u>…" This new edition, besides minor corrections, is highly enriched by clinical case presentations you will enjoy.

Maya Publications
P.O. Box 200-234
Bay Ridge Station
Brooklyn, NY 11220

For further information or questions, please write to the above address.

The information in this book is not intended to be used for medical or psychiatric diagnosis of any person. The illustrations included are approximations of the subject matter for ease of understanding.

I would like to thank Nancy Lopez for her support and assistance with the final completion of this book. I would also like to thank my friend, Casto F. Rodriguez Arbelo, who is a lawyer and a poet. Your contribution to the development of our community in Brooklyn is very much appreciated.

Table of Contents
(This is an incomplete listing of the topics covered in this book.)

THE DNA DOUBLE HELIX

The date was February 28, 1953. There were no news coming from the Korean War that excited James D. Watson brain cells. The day before, chemist J. Donahue from California Institute of Technology made a significant correction to the nuclei acid textbook Watson was relying on to construct a structural model for the DNA macro-molecule. How many chains is the structure composed of: two, three or more? Where would the basis adenine, thymine, cytosine and guanine go? If I place them on the outside hanging from the DNA backbone, can they come together to replicate themselves creating a genetic code for all animals and plants? How can the basis A, T, C. and G, be bonded to each other and pass on the genetic message to future generations? Which basis will attract, or repel each other? How do we go around testing our hypothesis? The question whether protein or DNA was the hereditary material had been scientifically established at Cold Spring Harbor in New York by Alfred Hershey and Martha Chase. American and British scientists seemed to have been engaged in a competing race to discover the structure of the book of life. How was this book written? Was it written in a code or it randomly came together?   Was the code a chemical or an electric one?

There were brilliant and outstanding chemists, physicists and biologists on both sides of the Atlantic Ocean very much interested in covering the first page of scientific journals and major newspapers around the world. In the U.S. A. we had world re-known, twice Nobel Prize winner chemist, Linus Pauling. He was very much engaged in the race to bring to America the Nobel trophy. In England there was Rosalind Franklin, a pioneer X- ray crystallographer who was very close to discovering the double helix DNA model. Not too far from Rosalind work place was Maurice Wilkins at the Biophysics laboratory of Kings College. These two brilliant English scientists were working close to each other. However, their own bodily chemistry kept them miles apart, except when Watson and Crick, but especially, when Watson came around trying to push his double helix model.

James D. Watson, our American biologist was desperately trying to find his place among the most prominent scientists of his time. His traveling between America and Europe put him in contact with influential scientists and scientific organizations on both sides of the Ocean.

His scientific curiosity, approachable personality and perseverance landed him in a research position at the Cavendish Laboratory at Cambridge University in the United Kingdom. Our maverick and future Nobel Prize Laureate, J.D. Watson was blessed by sharing space in the biochemistry laboratory with a WWII, twelve years his senior, physicist, Francis Crick working toward his Ph. D. in biology. Watson and Crick, an American and a Britton, must have had each other's chemistry well lined up for mutual compatible communication. Francis Crick seemed to have a pleasant, happy, gregarious and easily contagious mood and personality. In many ways, he was the perfect complement to Watson. Crick not only came out with brilliant ideas of his own, but for the good of both, Crick was always carefully attentive to Watson observations when listening to his colleagues or attending lectures and conferences by other scientists. When anything did not go right, it was expected, at Cavendish, for Cricks` good humor to turn it into an insightful moment to celebrate with laughing jokes and a favorable spirit to move forward with their project.

On the other hand, you had two hard headed Brittons, each one following their own research goals and thinking. Rosalind Franklin had many years of formal training and experience over both, Crick and Watson, and she wanted test-results on crystallography diffraction before making any premature announcements. Cause and effect as well as test- result were her favorite criteria for claiming a job well done. The other key player in the double helix was physicist Maurice Wilkins whose past experience included working on the Manhattan Project. Whether he met and shared his work with Albert Einstein and Robert Oppenheimer, we do not know. However, we assume that Wilkins was a good team player for him to survive the intrigues and secrecy of his job. Despite professional intrigues, he maintained friendly communication with R. Franklin and cooperated with Watson and Crick while working on the double helix. As already mentioned, we in America had a super star chemist, L. Pauling at

Caltech, dictating how chemistry should be read and practiced. Hardly anyone would challenge his arguments as he had discovered the exact arrangement or sequence on how amino-acids fold up into proteins, the building blocks of our body. However, on the DNA double helix structure, he missed the point by a very small fraction, but still a fraction. Otherwise, he would have been a three times Nobel Laureate. Wilkins, the physicist, may have had much in common with Rose Franklin on research principles; but socially, they must have been worlds apart. As J. D. Watson has put it, when the Lab. day was over, Rosalind dropped her working apron-gown and put on a stylish and distinctive evening dress and mingled with London's elite society. We add that it has never been proven that we are all equal or that we enjoy eating the same apple twice.

However, it seems that neither Watson nor Crick would be intimidated by occasional road blocks impeding or delaying their search for the DNA molecule structure. The United Kingdom Cambridge University was a gold mine of information on DNA. Both scientists complied with their superiors' orders to work on proteins at the expense of the double helix model. While following orders, working on proteins, their ultimate goal, the DNA molecule structure, never left their brains. Watson's eyes caught a scientific paper reporting that his much looked for DNA molecule bases, adenine and thymine as well as guanine and cytosine, appeared in roughly equal amounts. Watson's brain was well focused grasping this chemical balance inside his golden DNA molecule. But, it was his ever listening friend and colleague, Francis Crick, who foresaw the beginning of the solution for the problem: the chemical attraction of each base pair toward each other. Adenine would attach itself only to thymine and guanine to cytosine. Two chains, each one holding opposite bases, and chemically attracting each other could explain the structural model of the double helix and subsequent duplication. Rosalind Franklin X-ray photos of the molecule were a decisive factor achieving this historic moment in human biology. On February 28, 1953, the key features of the DNA model all fell into place. The two chains were held together by strong hydrogen bonds between adenine-thymine and guanine-cytosine base pairs.[1] It all fell in place almost as Rosalind had seen it during her visit to Watson and Crick laboratory room. Subsequently, they enjoyed the Nobel Prize; unfortunately, Rosalind had died of cancer before the award was given. No Nobel Award is given posthumous.

April 25, 1953

On April 25, 1953, the world prestigious science journal, Nature, published a one page article with the title: Molecular Structure of Nucleic Acids- A structure for Deoxyribose Nucleic Acid, signed, J.D. Watson, F.H.C. Crick

There it was, a single page showing the drawing of this double helix with its sugar-phosphate backbone on the outside in a twisted fashion holding inside the nucleic acids. How tempting it was. Was it possible that I had the book of life in front of me? The authors had written, almost at the end of the article: "It has not escaped our notice that the specific pairing we have postulated suggests a possible copying mechanism for the genetic material." Is this a possible mechanism for copying the genetic material? Did it mean that there is a genetic code for me to find out how I was made? The basic elements comprising my body, including protein structure, was already shown to us by L. Pauling. Life, as it had been taught for centuries, took a different turn with this single page. From now on, the main focus of Biology was to understand the working dynamics of the double helix and its nucleic acids forming it. That single page provoked more questions than any other discovery in the XX century, according to most biologists and scientists in general. The world had had hardly time to digest and accommodate itself to Darwin's, On the Origin of Species, not to be frightened by a code of letters, A.T.C. G. to open the alphabet of our book of life.

How did the double strands separate from each other for self-replication? How and where are proteins formed? Is the code for proteins synthesis implied in the four letters? Francis Crick had postulated the dogma: DNA—RNA –Proteins. How does it relate to each other? Is the nucleus of the cell the command post for all

---

[1] - James D. Watson DNA, Alfred A. Knopp, New York, 2003, p. 52

cellular activities? Where in the cell are the letters synthesized? How many letters are there in the book of life? Watson and Crick in one page article on 4/25/53 in the journal, Nature, put biologists, chemists and physicists to do science 24/7 even during their dreams.

Protein synthesis kept scientists busy until they discovered ribosomes as the assembly site for proteins. It was Paul Zamecnik and his team at Massachusetts General Hospital who further elucidated for us the role of R N A in protein formation. They confirmed Crick's adaptation theory naming it transfer RNA. This molecule hauls amino-acids groups that would exactly match another RNA strand coming from the nucleus of the cell, dubbed messenger RNA. The year was 1960. Messenger RNA was proven to be the true template for protein synthesis. We had not only discovered the letters of life alphabet, but began to learn how the acids of life work. Three world re-known American research centers, Harvard, Caltech and Cambridge, Massachusetts had done the work. M. Messelson, F. Jacobs and S. Brenner, demonstrated ribosomes existing outside the cell nucleus were, in effect, protein assembly factories. There are many ribosomes inside the cell. They look like tiny beads set in line in the cell's cytoplasm. The inside of the cell was becoming clear for scientists looking for answers in Biology. When they found different organelle's structure and functions, it was exciting news for scientists and the general public. We were working with the human genome and we were excited with the discoveries of this new science, molecular biology. Our enthusiasm prompted some individuals to make premature announcements. The cell inner organelles, their shapes and functions were under constant microscopic scrutiny and testing. Nobel Prize in Chemistry Laureate, Ada Etil Yonath, who participated in a meeting of the World Chemistry Congress in Puerto Rico, said that 60 percent of the ribosome is RNA. She added that RNA is a machine from before life evolved. Life grew around it. RNA was one of the first molecules.[2]

---

[2] - Nature, 10/13/, 2011, p.56-57.

Unresolved as of yet there was a nagging question. How did nucleic acids, those repeating letters: A.T.C.G., become peptide chains? In other words, there are four letters and twenty amino-acids. Which and how many amino-acids would fit into a letter to form a peptide chain? Once more, it was Francis Crick and his colleague, Sydney Brenner in 1961 at Cambridge University who proved that the code was a codon, also known as a triplet.

Three letters for each amino-acid was the solution to the problem. However, the genetic code may come in more than one codon for each amino-acid. For example, the amino-acid tryptophan- comes in a UGG codon only, while arginine and serine may come out in six different triplets or codons. Word of caution, in the DNA sequence we have thymine which binds with adenine. In RNA, it is uracil that complements adenine. Marshall Nirenberg working at the U.S. National Institute of Health participated at an International Congress in Biology in Moscow in 1961. Nirenberg was a young scientist hardly known outside his close friends. He was allowed to speak for about ten minutes. Like during any other event, you like to hear the most important personalities in the field, not an unknown neophyte. Even our J.D. Watson was chatting outside the conference room. Nirenberg reported that in his Lab, he had gotten a ribosome to pump out a simple protein known as phenylalanine that comes out in codons UUU or UUC. Once more, it was Crick who arranged for Nirenberg to speak to a hall packed with every scientist that could get to Moscow the following morning. He was later awarded the Noble Prize. Nirenberg had put everyone in the run to finish the genetic code for each amino-acid.

F. Crick dogma: DNA-RNA Proteins seemed to have been well established. From the cell nucleus, DNA is transcribed into a messenger RNA molecule that goes to a ribosome in the cytoplasm. Ribosomes can be seen as tiny beads (like beads in a necklace) in the cell away from the cell's nucleus. Another RNA molecule named transfer RNA delivers amino-acids to ribosomes for the synthesis of proteins. Each ribosome acting as an assembly factory, paste together messenger RNA and transfer RNA to form or build proteins. It is helpful to remember that replication exist at the double helix un-zipping process while transcription takes place while the cell nucleus is forming single RNA strands. Another term frequently used by most scientists when talking about these processes is translation. Translation is basically a process that takes place at ribosomes locations in the cytoplasm. Each molecular strand carries a set of letters that complement each other. For example, a messenger RNA triplet CAA will recognize and chemically attract the corresponding triplet GUU in the transfer RNA. Ribosomes will be in charge of pasting together these two molecules, thus forming peptide chains, known to us as proteins. Its shape will determine function and some peptide chains are called enzymes. Enzymes are very busy catalyst agent in our body. They are always provoking chemical changes without we ever knowing anything about it. As soon as I put a teaspoon full of ice-cream in my mouth, enzymes begin to do their work. Enzymes do their work inside the cell as well outside around the cell. We will address the ribonucleic acid or RNA later on in this essay because strands of this acid seem to be the working horse in protein forming and research tools. Recently, the traditional Watson-Crick dogma for protein synthesis has been under review.

We will be going back and forth in this essay trying to put together Watson and Crick milestone scientific success story. After Watson and Crick had gotten the double helix DNA structure, let us say, approved by the majority of scientists in the field, one of the questions popping up in my brain all the times was: Where and how are the building blocks of our body (protein) synthesized? What roles do the ribonucleic acid plays in life forming? Does it compete with DNA in protein synthesis? How are ribonucleic acids involved in the complex process of peptide chains? We already advanced you some of these answers, but at the time, it was mind bugging for all involved; especially, for Watson and Crick. It was the good humored F. Crick who once again suggested the dogma: DNA—RNA—Proteins. DNA polymerase, the enzymes that unzips the double helix, was well established. Actually, it is protein helicase that unzips it. Soon, RNA polymerase (an enzyme) came to take its corresponding place along with DNA polymerase. The nucleus of the cell was the commanding center sending out messages and ordering cell's organelles what to do. However, it was RNA multiple forms that moved around strands to assemble proteins at ribosomes` sites. The template or mold for protein synthesis, dubbed messenger RNA, came out of the nucleus; but it was transported outside the nucleus into the cytoplasm to be matched with another RNA molecule carrying an amino acid destined for a ribosome. As we indicated above, translation takes place at this level of protein synthesis. It is also the place or process where most problems take place.

Translation, whether it is chemically matching nucleic acids or language translation, it seems to create headaches for many of us. It is hard to find and match words or phrases that exactly convey the same meaning and feelings in any two languages. Translation in our work is an issue of RNA molecules. During the replication process, the DNA polymerase (helicase enzyme) unzips the double base pairs chain and it will re-create itself instantaneously by adding the necessary complementary nucleotides to each strand. The original strand of letters, for example, A for adenine, T for thymine, G for guanine and C for cytosine will be followed by more letters-nucleotides- in a sugar-phosphate backbone running in the opposite direction, thus forming a double strand. The letters or nucleotides will chemically attract each other forming strong covalent bonds. They share electrons. You may have a strand composed of T.G.A.C.G.T.T.A.G., all these bases held together or attached to a sugar-phosphate bone. In this hypothetical case, the complementary bases will be as follow: A.C.T.G.C.A.A.T.C. You have T and A bonded together followed by the rest of letters running anti-parallel to each other. This is DNA replication. During genome sequencing, if you know one strand, the accompanying strand is not hard to guess. Your genes, the genetic material inside your cells, as well as every cell in all living things on planet Earth, are made of DNA. Each strand of DNA consisting of many poly-nucleotides is bonded together by hydrogen bonds. Although for illustration purpose we tend to draw and describe each DNA strand in a straight and bar-like structure, but in reality, it is a three dimension structure in which each strand coils around each other in a spiral fashion. Most likely, you see the length and width of the strand; but the depth in most books description, needs your imagination to be completed. We want to clarify that the word nucleotide refers not only to the four nucleic acids bases A.T.G. C. Generally, it is used to designate the bases with their sugar-phosphate group attached to it. Our DNA is well protected inside the nucleus of the cell. It is a tightly controlled structure. It needs to be wrapped that way because it needs to protect itself from invaders inside the cell cytoplasm as well as extra-cellular intrusion and interference.

Recently, the field of genetics as well as out of the womb causal factors has pointed-out risks and threats to genes and possible diseases. Before we proceed with this essay on nucleic acids, we like to remind you that the twenty amino acids needed in our organism are polymers for the synthesis of proteins. They possess an amino acid group, namely NH_2 at one end and a carboxyl acid group, COOH, at its other end. Both groups are bonded to a single carbon atom known to us as the a-carbon. There is a side chain linked to the a-carbon altering the shape of the molecule, thus changing its chemical function.[3] There are more than twenty amino acids found in nature, but only twenty are needed in your body. They repeat themselves over and over again in proteins in bacteria, plants or in a privilege civilized human being like you and I. Why nature chose only twenty amino acids to be the building blocks of protein is another question to be answered in the future. Also, there are about 92 elements that commonly appear in nature, but less than a third is used by life on Earth. You will enjoy our trip into our body endless action and re-actions. It is a marvelous and sophisticated chemical engine. Have you wonder how D.N.A and R.N.A paired together? How about mitochondria, they are organelles outside the nucleus, were they captured by the nucleus to do its work too?

---

[3]- Essential Cell Biology, Alberts et al. Garland publishing Inc. New York, 1998, p. 60.

Following replication, we have transcription. With the help of RNA polymerase, the DNA molecule transcribes- meaning, forms messenger RNA, transfer RNA, interference RNA and other ribonucleic acids strands. For the sake of clarity, when you replicate (copy) a page or paragraph of a book into a home-work assignment, most probably, it will come out without any mistake or error. But try to translate the same page from English to French, Spanish or Chinese, and in all probability, your professor or teacher will find room for improvement. If you are translating poetry, it might prove even harder to make perfect translation. As I said earlier, translation is an issue of RNA molecules. RNA is not a faithful translator for reproductions of DNA genetic instructions. Watson, Crick and colleagues established the dogma that the chemical letters A.T.C.G. are the genetic code while the cell's work horse, the RNA, is made in its likeness with a minor exception, the uracil nucleotide. RNA, the nagging acid, was loaded with the responsibility of transporting molecules to ribosomes for the synthesis of proteins. These proteins are used during the making of multi-cellular organisms like a giraffe, an elephant, a turtle and a human being. We wonder how many misspelling errors are found in each genome of these animals. Researcher Mingiao Li, a geneticist at Penn Medical School in Philadelphia conducted a research study of RNA in which he included blood cells from 27 different persons. On average, each person has nearly 4,000 genes with RNA containing misspelling not found in DNA; while RNA has more than 20, 000 different places in the genome. The most common misspelling changes occur in the A letter in DNA which is changed to G in the RNA strand.[4] The researcher took precautions that variables, including a virus did not invalidate or stained his work. We must not forget that the world of RNA is not only very old, but abundant. While DNA sits in the nucleus of the cell giving out orders and sending out messenger RNA for the synthesis of protein, ribonucleic acids molecules are busy transporting and delivering polymers inside the cell. We already know that without messenger RNA, there would

---

[4] - Science News, 12/04.2010, Vol. 178, # 12, p.17

not be protein synthesis and, without RNA in the ribosomes, peptides would not be pumped out of this assembly factory. Besides, micro-RNA is being used as engineering tools to work not only inside the cell organelles, but to modify genes. Do not be exaggerated concerned by so many misspellings and changes of one letter or a base for another. There is a self-correcting system inside our macro-molecule DNA. Occasional changes of letters or bases are conducive to mutations, and mutations, without losing the basic core of the cell, made you and I beautiful and intelligent human beings. Ask your mother and she will re-assure you are the most beautiful and intelligence individual in the universe.

The first micro RNA

The first micro RNA was discovered in 1993; the second one was discovered at Massachusetts General Hospital in the year 2,000 by Gary Ruvkum. In between these two discoveries, in 1998, perhaps the most versatile ribonucleic acid molecule, interference RNA was added to the family. Among other things, it can interfere and suppress or inhibit a gene from forming proteins. We would like you to remember throughout this essay the potential power of t-RNAs as well as micro RNA's. The potential for use and abuse of this molecule is beginning to be understood. All genetic information encoded in the DNA of eukaryote (possessing a nucleus) cell genes is transcribed by RNA polymerases. The process of transcription is dominated by proteins, but a growing number of them are RNA.[5] Recently, a new RNA molecule dubbed e RNA (enhancer r n a) was discovered. It belongs to non-coding RNA group. Enhancers are engaged in activating the transcription process. It is suspected that an enhancer RNA contributes to human disease and the evolution of human specific traits.[6] We need not to name for you specific human traits. What makes you different from a chimpanzee besides the color of your eyes, skin, language and intelligence to make wise decisions? Besides the problem encountered by the translation process and epigenetics, we would like to briefly mention the so called "junk

---

[5] - Nature, 09/10/2010, Vol. 461, # 7261, p. 186.
[6] - Nature, 05/13, 2010, Vol. 465, p. 173.

DNA" also known by non- coding DNA. It has been the subject of much debate and scrutiny over its usefulness in our organism. If it does not code for proteins, what is the use of having an excess of cells that may be prone to health problems and diseases? Is this junk DNA a left over from our earliest evolutionary developmental process that needs to be modernized or recycled? Perhaps, non-coding DNA is like glia- cells, neglected by scientists because neurons, like coding DNA for proteins, has taken first place in research laboratories. An alarming 90 percent of our genome is composed of non-coding genes. A recent study done at Stanford University comparing human genetic blueprint genome with those of other animals concluded that very little of the human genome is really necessary. Andrew Sidow, the principal researcher, claims that about only 7 –seven percent of the human genome is similar to the DNA of other mammals. This researcher narrows down to 225 million of the 3 billion letters (A.T.C.G.) of our DNA.[7] Not all scientists in this field concur on this estimate. During the process of transcription, when a strand of messenger RNA is being made, the junk or non-coding genes are cut off, so the messenger RNA comes out clean and safe to be transported to a ribosome where it will be converted into a protein for your beautiful and well-shaped body.

Our interest in RNA

We have been interested in the RNA molecule because it takes multiple forms and in its translators` job, it is the subject of multiple problems that may contribute to protein dysfunctions and diseases. However, before we proceed on our tour of the genome and the genetic code, we are going to say a few things about the DNA macromolecule. DNA has the potential for change without losing the basic components of its ancestors. It allows evolutionary processes to adapt and adjust itself to new environment, but it maintains the bases of its genetic code. This genetic code has and

---

[7] - Science News, 12/04/ 2010, Vol. 178, #12, p. 77.

transmits information for the purpose of creating multiple different organisms. DNA establishes amino acids sequences for protein synthesis. This great molecule created a brain capable of studying and repairing itself by protein synthesis and memory storage. (RNA molecules are prominently involved during this process). DNA molecules provide for the continuity of the species with minimum changes over a long period of time; long time may mean millions of years. If we include the Neanderthal man as part of our human ancestry lineage as I do, we have been here for a long time. Recent discoveries by Svante Paabo from the Max Planck University in Germany, as well as discoveries at Atapuerca in Spain and southern Siberia in Russia, all point to the presence of intelligent human beings on dear mother Earth long time ago.

The DNA molecule has its own membrane within the cell cytoplasm that guarantees its integrity and identity as if it were the command center of the cell. From this complex DNA molecule comes out instructions to construct a beautiful humming bird, a mouse, an elephant and a human being. It has the information to replicate itself in double helix or just simple strands of RNA. Our genetic code, the four letters: A.T.C.G. is in itself the great DNA molecule. How these letters are attracted to each other is written in DNA. This molecule keeps in its nucleus the instruction for the color of your eyes, hair and skin. To some extent, it also determines how tall or short you will grow to be. Of course, environmental factors have input deciding your height and weight. A poor diet composed of simple carbohydrates without protein rich beans, grains like soy beans, green peas, lima beans, kidney beans, chick peas, and affordable tuna fish, sardines and salmon, does not need to be explained any further. Likewise, throwing down your throat fried pork chops, red medium cooked steaks seasoned by artificial flavors and excess salt; you can rest assured that it would be extremely hard for you to be a healthy person at age 90.

We have become interested in the ribonucleic acid molecule not only because it transforms itself in multiple forms, but because in its interference form can inhibit the formation of our own bodily tissues, including brain cells. The micro RNA hair pin molecule is also another relatively new discovery. Thanks to its small size, this molecule has been used to deliver research and therapeutic tools inside the cell. It can be used to engineer gene modification in a brain disease or any other bodily pathology. It has been used in lower animal models. Time, dedication and, of course, funding, will decide the future of all these new technologies. Micro RNA can be used to target genes that at a certain age of bodily development provoke a dysfunction or a specific disease. These diminutive molecules can take its place in amino acids chains and alter the completion and function of a faulty protein or gene. During the blastocyst phase of embryonic growth, it can be used to silence pathological cells. Gene silencing through RNA interference and similar micro RNAs has become the tool of choice for analysis of gene function, diseases and drug discovery. Among the smallest of these molecules besides the micro RNA, we have short interfering and hairpin screening tools which are first choice among genome researchers.

FIGHTING DISEASES

Our next challenge to Watson and Crick book of life is how the genetic code can be translated into fighting diseases and improving human health in general? From studies of mummies and animal fossils we have learned that some diseases have been haunting us for thousands of years. The study of nucleic acids- the base pair of letters- in our genome can be used to point out mutations, missing segments of letters, misspellings and repeating non coding genes. Why has it been so difficult to fight cancer, schizophrenia, Alzheimer and other degenerative diseases in our organism? We have engineered animal models for laboratory work like mice, worms, bacteria and fruit flies, among others. It has been a challenge for all of us who would like to see human suffering alleviated with newly found discoveries, mainly after WW11. Privately funded laboratory researchers, pharmaceutical companies, State and Federal agencies have joined efforts to find a solution as soon as possible. During these intervening years much funding was diverted to space exploration in competition with the Soviet Union. The exploration of human somatic and brain cells was lagging behind. We did not have the technology-the working tools- to observe, catalogue, categorize and analyze the behavior and role of each DNA molecule. The same work had to be done with ribonucleic acid molecules. Whatever little work was done, it was done by hands! Although the double helix and both nucleic acids were known in general, the dynamics, physiology and chemistry of the brain was left to Sigmund Freud, C.C. Jung, theologians and poets to speculate about it. For two thousand years, our cultural and religious heritage from Greek and Roman philosophers had speculated about our illness, especially, mental diseases. The old Greek scientific knowledge was buried under superstition and religious fanatics during the middle ages. It was not until the Renaissance that scholars from Byzantium introduced it to Western Europe.

A mouse brain tissue

The slicing of a mouse brain tissue, observing how a cell grows in a Petri dish or a virus multiplying in a bacterium environment did not receive the necessary support from the public. For gene sequence history lovers who enjoy themselves engaging in mind traveling, we like to remind you that by 1985, the longest genome that had been sequenced was a virus of 172,282 base pairs. The human genome has a little over three billions. Do you understand what we mean? We were sleeping, and sleeping soundly. That was 32 years after Watson and Crick had given us the genetic code of life. Three worldwide known leaders and researchers took over the audacious if not gargantuan proposal: SEQUENCING THE HUMAN GENOME. We were still guessing or speculating about how many genes were there in the genome. The three visionary leaders we are referring to are, James D. Watson, Francis Collins and Craig Venter of the Celera Genomics. As late as 1988, James D. Watson was named associate director of the Office of Genome Research, part of U.S.A. National Institute of Health. He was in charge of the genome project until 1992. Francis Collins took over from Watson until its sequence was completed in 2003. Craig Venter published his human genome sequence in the journal Science while the N.I. H genome was published in Nature, a British journal. By the time Watson, Collins

21

and Venter took upon themselves the sequencing of the human DNA, the tool that multiplies DNA fragments for analysis, PCR was available to them. Making use of restriction enzyme to splice or cut DNA strands at target sites became very useful tools not only for reading the sequence, but to mark genes involved in inheritance diseases. Some of the earliest molecular biologist and scientists working with DNA structure and its content claimed that genes made up only three percent of DNA in human cells. That claim was made in 1995. Today, we call non-coding DNA, introns while coding DNA is dubbed exons. As more sophisticated research tools are developed, the so called junk DNA become under meticulous scrutiny and analysis. These non- coding genes are revealing their usefulness. The race to be the first one to sequence and read the book of life was centered in America with minor international contributions. However, as government and scientists around the world became aware of its potential benefit not only as a wonder tool fighting disease; but also, as an agricultural aid improving food supply to the world population, networks were established for joint research projects. Seeds can be engineer to improve harvesting seeds, accelerate animal and chicken growth. Genetic engineering can be the tool par-excellence to feed hungry people around the world where climate and weather disasters make it difficult to grow enough food for daily consumption.

The GWA

For this purpose, Genome Wide Association- G.W.A- studies have engaged hundreds of individuals, both researchers and patients, identifying disease markers in their DNA. Deciphering the book of life language has arisen- wake up- the mind of many brilliant and concerned persons on planet Earth. Using the advances of Restriction Enzymes and Recombinant DNA, we have moved forward in our attempt to bring relief to the suffering and hopeless individuals at worldwide level. We are using genes to help patients, as in gene therapy. We have been using harnessed virus as a vector to deliver therapeutic tools to target areas. The usefulness of the human genome sequence benefits is not in question. Many individuals have hailed it as the greatest human achievement ever made in human history. Of course, there are alarmists people who have religious, philosophical or superstition claims that interpret man's advances a challenge to their world view. We are going to enter two quotations from two different professors addressing Craig Venter latest achievement with a synthetic bacterium. "A prosthetic genome hasten the day when life forms can be made entirely from non-living materials. Another professor wrote: "Venter's achievement would seem to extinguish the argument that life requires a special force or power to exist. This makes it one of the most important scientific achievements in the history of mankind." [8] Following that, I would like you to ponder on the quotation by another professor of biomedical engineering who appeared on page 424 of the same journal. "The micro-organism reported by Venter's team is synthetic in the sense that its DNA is synthesized, not that a new life has been created. Its genome is a stitched copy of the DNA of an organism that existed in nature, with a few small tweaks thrown in." You have the opinion of three different professors, all reading the same article in the same journal. One praises it, another puts down while another sees it as an assault to almost everything we believe in. When C. Venter appeared on T.V. announcing the creation of a synthetic bacterium, we do not recall him claiming the forming of new life. We understood it as a brilliant genetic engineering accomplishment of mankind never seen or done before. He had opened the door for a new approach to make genes and proteins work for us at our own level of development. Venter`s bacterium could be turned to work for us manufacturing chemicals for our brain's dysfunctional organs. In Parkinson disease, Venter's bacterium could be put to build dopamine, replacing the substantia nigra dopamine deficit. I can see it adopted in agriculture engineered to kill bugs instead of health risk pesticides. A bacterium engineering could be used to increase food production to feed our world population. We have many variant genes that seem to be idle or not performing at our own level of self-development. Venter has proven that we have grown up; he has proven our humanity. We can take charge of our own organism.

---

[8]- Nature, 05/272010, Vol. 465, #7297, p. 422-423

He just finished sequencing his own genome, practically a few seconds ago in our life history.

C. Venter challenged HGP

C, Venter had challenged the Federal Government sponsored Human Genome Project and came out a big winner. His success relied upon improving existing technology to assemble and improve on nature's already existing life structure. Life itself is dependent on little beats of elements coming together and functioning as a separate unit. These units of life are around us, and the universe is full of it. C. Venter's concept of throwing in little pieces of life elements (genetic engineering) into the cell was around for almost four decades. We are referring to recombinant DNA pioneers, Herb Boyer and Stanley Cohen's work. Without hesitation, we can call them the first and most accomplished genetic engineers. The beauty of C. Venter's team creature, the synthetic bacterium is its ability to engineer organic material from the universe (which our planet is part of) to assist and improve human evolution. When we first cut and isolated a DNA molecule, making use of restriction enzymes thus putting our fingers unto genes, we continued our game with our alphabet of life throwing in the ligaze enzyme and bingo, it glued together and a plasmid came out to exist as another tool. These two new tools of research and therapy, restriction enzyme and recombinant DNA, took over the front pages of scientific journals, newspapers and newscasts on T.V. From building and engineering a bacterium, we can go on and rewire naturally existing endogenous gene circuits and pathways. I am aware that this would a difficult but gigantic step forward. However, Zhen Xie et al, a Harvard University professor, reported a discovery that appears to be worthy of further study and analysis. His team has developed tools we can learn to use at present time. Incidentally, George Church's bacterium engineering might come handy producing killer cells (apoptosis) without harming normal and healthy cells. Cancer cells are terrorist cells. We should be channeling funds to identify and destroy terrorist's producing cancerous cells. C, Venter, G. Church, Zhen Xie, along with other pioneers in this field area, are ahead on this race of technological revolution. We need to transfer and apply this technology into practical use for anyone who needs it.

When we were cloning DNA sequences, cutting with restriction enzymes and linking them together with a DNA enzyme named ligaze, we were throwing in "little tweaks" in the genome cells. We were using the bacterium cell as the working horse as it divides over and over again in a culture. Boyer and Cohen were using an existing micro-organism for the benefit of a multi cellular organism. We can argue that Boyer, Cohen, Venter and other scientists are just accelerating and improving on the evolutionary work begun many millions of years ago by eukaryote organisms. The evolution of life on Earth has had multiple twists and turns not fully clear to most of us. One of those twists must have been when the RNA molecule made advances to another powerful and interesting molecule, the deoxyribonucleic acid molecule, DNA.

Francis Crick and RNA

Our friend and brilliant scientist, Francis Crick, according J.D. Watson, as early as 1968, suggested that RNA must have been the first genetic molecule..."besides acting as a template, might also act as an enzyme catalyzing its own self replication."[9] This insight of Crick on the ribonucleic acid when the double helix and its nucleotides were still trying to find their rightful place in our brains, was awesome. This fellow seems to pull his ideas out of his jacket's rear pockets, and most of his ideas were proven right. Even on his way to the hospital before his death, he was doing science in the ambulance, according to some of his biographers. We are grateful we have guys like him working for humanity.

A team of researchers from one of the most prestigious Universities in the U.S.A. wrote: "The reality of R.N.A., an R.N.A. dominated stage in the early evolution of life prior to the evolution of coded protein synthesis, has been firmly established by recent structural studies of ribosomes.[10] We must point out that in the protein assembly factory, a ribosome, one of its components is

---

[9]- The R. N.A. World, 3rd edition, Cold Spring Harbor Laboratory Press, N.Y., 2006, p. XXIII.
[10]- Op. cit., p. 57.

ribonucleic acid. This acid in its multiple molecules is not only an effective working horse. In addition, in the ribosome, it is part of the machine that builds very complex organisms like human beings. We will continue to address the chemical composition of a ribosome. Further elucidating cellular and life complexity on planet Earth, we must attempt to answer, how did the mitochondria- the cellular energy engine- come to be part of our organism's cells? Nature has been working with and engineering our basic functional units, our cells, from the very beginning of life itself. All our basic elements beginning with carbon, nitrogen, hydrogen, oxygen, potassium, sodium, etc., are part of nature. Watson, Crick, Cajal, Collins, Cohen, Boyer, Venter and all of us, are just nature in its natural evolutionary development. It seems that they are among nature's most developed creatures. The contribution by pioneer scientists sequencing the human genome has helped our relationship with other species. It has helped us understand ourselves by understanding our closest hominid genome, the chimpanzee. The mouse, not too far behind, is among the best research tools we have. By engineering and observing human diseases in a mouse, we can trace genes responsible for the disease and develop corrective therapeutic tools.

Current worldwide issues like climate change may also benefit from hominid genome sequencing. Anthropologists, historians and scientists of multiple orientations must have enjoyed Svante Paabo's work at the Max Planck Institute in Germany. His latest discovery is on Homo sapiens traveling around our planet. He brought up new issue with the "out of Africa" theory as part of mankind history on our planet. He posed serious questions about climate changes and their impact on plant and animal survival on Earth. Neanderthal once occupied most of Europe from the Iberian Peninsula to southern Siberia in Russia. Climate change brought drastic weather changes and most of the continent was covered by heavy snow all year around. How much did glaciers contribute to Neanderthal's extinction needs further study and analysis. This study relates to us in terms of demographic movement around the world and its impact on us. The sequence of human genome from different parts of the world's population helps us understand how genetic and environment participate and influence each other in disease etiology. Which genetic factors are basically biological and which are environmental factors? Through worldwide genome studies we can identify the most common variants from samples in China, Japan, India, Russia, Europe and U.S.A. For example, does autism appear in the same gene and chromosome in children in any of those countries? Biomedical technology developed in the U.S.A. using multiple RNA molecules tools can be used to bring the East and West together in fighting diseases and plagues in animals and plants. It seems we trade with every country in this planet. We eat their food stuff and handle many of their home made products that may provoke spontaneous diseases.

Beyond the HGP completion in …, what

Following the completion of the Human Genome Project in 2003, many surprises linked to gene's structure in the chromosome have come to light. Large segments of non-coding genes are present far away from genes associated with a disease, however, it looks that they are linked to a disease. In some cases, gene variants common to autism have been linked to brain diseases like schizophrenia and manic depressive illness, also known as bipolar depression. Autism, like schizophrenia, is a multiple symptom brain disorder. Multiple genes and environmental causal factors may be involved in triggering its behavior. A group of international researchers studied nearly a thousand autistic persons and compared it with an equal number of healthy individuals. They found dozens of genes involved in autistic disorder. Interestingly and rightfully surprised was the fact that most of the variants in autistic disease were missing segments or duplicated segments of DNA. Some of the genetic changes were inherited while other have to be attributed to parents or the child himself (epigenetics).[11] Outstanding symptoms of autistic children are their severe social limitations or difficulties, besides language problems and repetitive behaviors. We will include the following research finding of 2006 because it covers a large number of families with autistic members. "The gene dubbed MET regulates production of a protein that influences cell proliferation in various parts of the body. This gene lies on a stretch of chromosome 7 that other researchers had linked

---

[11]- Science News, 10/23/2010, p.18-21

to autism. Neurologist Pat Levitt from Vanderbilt University, the team leader, consulted database from 204 families with one or more children with autism. Another group of 539 families was included in the study and found that the link between the Met variant and autism appeared primarily in families with two or more affected children. Laboratory tests showed that this MET form lowers the gene's activity and reduces its production of proteins that binds to various tissues. Daniel H. Geschwind from the University of California at Los Angeles praised the report as the first time someone had identified a candidate gene for autism; however, he cautioned that Met could be the tip of the genetic iceberg. [12] A few years have passed since Pat Levitt conducted and published this report. As quote number 10 indicates, more recent research done by scientists attempting to untangle the genetic roots of autism, have found dozens of genes involved in this brain disease. But, we emphasize once more, there is more than just genes implicated in this disease. Autism, like schizophrenia, is not limited to race, country, ethnicity or social status. Common efforts to find a cure might not be too far away.

Autism affects mostly boys. Finding out the etiology of a genetic disease is not just pointing out gene duplication, missing segments of the genome sequence, molecular intrusion such as short nucleotide polyphormism (SNP) or a mountain of non-coding genes. Looking for rare variations in human genome is only one approach to multiple rare syndromes like schizophrenia and autism. What caused genome segments to be missing or duplicated in the first place? What provokes SNPs to appear in harmful locations? Researchers have shown that not everything is inherited. How about the embryonic phase of gestation? Parental mental and physical state at the time of conception undoubtedly contributes to embryonic, fetus, childhood and adults problems. The human embryo is subject to multiple risks and threats from inside as well as outside the mother's womb. Inside is not only a gene mutation by a letter change like A replaced by G. You have to consider epigenetics, a set of chemical alterations not found in the DNA code that partially manipulates gene expression. Among epigenetic markers we have methylation (CH_3), acetylation, phosphorylation

---

[12] - Science News, Vol.170, 10/21/2006, p.259.

and similar markers that may attach themselves to tails on histones and alter critical transcription processes. Histones are proteins that DNA coils around. Further complicating our job, we have micro RNA that may bind itself to a messenger RNA and suppress the synthesis of a particular protein. Suppose the protein that was not synthesized to its shape and remained incomplete or modified in some way and could not take its rightful place in a visual pathway in the occipital lobe? Definitely, it would put our vision in jeopardy. In this case scenario, outside environmental threats and risks has to be considered secondary to biological causal factors. A biological culprit can be found in any other vital organ of our body and not necessarily a dysfunctional gene. Revisiting Allis' hypothesis on histones, DiTacchio et al [13] showed us that the circadian clock that governs diurnal rhythms of physiology and behavior uses Allis' histone code extensively. However, Di Tacchio now showed that histone demythylase is important for the activity of the circadian activator complex, CLOCk-BMAL, and for circadian oscillation in general. In addition to Allis' hypothesis, modified histones serve as platforms for adenosine 5' triphosphate, ATP for short- dependent motor complexes that recognizes chromantin.

DNA packing

DNA packing could become an area of extreme care susceptible to harmful epigenetic intervention. Histones are proteins responsible for the first level of packing in eukaryote (cells with a nucleus) chromatin. The complex comprising DNA and proteins (histones) that forms a chromosome is called chromatin. The amount of histones in chromatin is approximately equal to the amount of DNA. Histones have a high proportion of positively charged amino acids (lysine, arginine) and bind tightly to the negatively charged DNA, thus forming the tightly packed chromatin. This tightly packed complex is a safeguard measure against outside intruders. It can be seen as nature's attempt to keep DNA uncontaminated and complete as possible. Seen under an electronic microscope, unfolded chromatin appears like beads of a

---

[13] - Science, 09/30/ 2011, p.1833.

string or necklace. For those with a religious orientation, it looks like beads in a rosary. Each bead is called a nucleosome. As you may have already suspected, chromatin organization in a nucleosome may influence gene expression by limiting the access of transcription proteins to DNA.[14]

In summary, chromatin is the molecular substance in all eukaryote cells that facilitates DNA packing efficiency. It protects itself from unnecessary un-raveling, unwinding itself and exposing its nucleic acids: adenine, thymine, guanine and cytosine to risky alterations. The DNA winds up, meaning, coils itself around histones, which are proteins which look like tiny beads. They make good binding to each other through negative and positive charges. However, for reasons we do not understand at present time, genome development through millions of years, allowed histones to grow tiny tails coming out of its beads like structure. These tails, like any external protuberance, may serve well the molecule, but the tails can be a tempting target for SNPs and other intruders to anchor and provoke havoc to our life creating acid, DNA. On the other hand, histone tails may serve as anchor sites for extra-cellular transporter of nutrients for both, DNA and histones. Besides growth factors, it may also serve as a listening post for communication among cells in bodily organs, and between neurons in the brain. How the cell keeps a balance between growth factors and SNPs while conserving for itself continuity-- the core genes that will keep the species genotype and phenotype with minor changes-- is another question we have not been able to answer in a satisfactory fashion.

A seemingly contradiction

It seems a contradiction for us that histones' proteins around which DNA is tightly wrapped allows tiny tails that may threaten its own structure and function. Cells are chemical factories going through multiple processes to keep its intrinsic functions. The cell has many channels in its membrane, including the energy

---

[14]- Biology, Neil A. Campbell, Benjamin Cummings Publishing Co. Ca. 1 987, p. 372-73.

producing dynamic molecule ATP (adenosine triphosphate) that plays a very significant role in metabolic processes. Does it have a role, a position in protecting the chromosome complex, in particular, histones' tails from undesirable intruders? Multiple proteins are involved in forming the chromatin complex, including the critical transcription and translation processes that are critically important for protein synthesis. We have mentioned before that multiple mistakes of different sorts take place during these two stages of protein synthesis. Proteins are crucial for DNA tightly coiling-wrapping around and shape forming, besides being part of ribosomes. A group of researchers interested, among other things, in DNA and histone's packaging in chromatin, studied an enzyme dubbed, 1sw2. It is an ATP-dependent chromatin remodeling protein highly evolutionary conserved. They found that the positioning of thousands of nucleosomes adjacent to important regulatory sequences is controlled by 1sw2. This enzyme is able to use energy from ATP hydrolysis to override the inherent nucleosome-positioning signal of the underlying DNA. 1sw2 may function generally to reposition nucleosomes on unfavorable DNA sequences. The ability of proteins such as 1sw2 to reposition nucleosomes, provides a clear illustration that cellular factors activity operate to disrupt the intrinsic cues that would otherwise package the genome.[15]

We hope to have given you an idea, although incomplete, about multiple problems we encounter just trying to identify possible loci for a disease. As mentioned earlier, it is not just carrying the genes of one parent or the other, but epigenetic and everything around us. Continuing with epigenetics, Frances Champagne, a neuroscientist at Columbia University in New York City, has said that DNA methylation is the most enduring of epigenetic modifications. It seems to be a complicated system of DNA synthesis, packing, replication and transcription. In a base pair, guanine-cytosine, the methyl molecule, CH_3, may just sit itself next to cytosine nucleotide in the double helix and become methylated. The problem is that methylated cytosine looks like another nucleotide called thymine and may be converted from

---

[15]- Nature, 12/13/07, Vol. 450, #7172, p. 1031-35

cytosine to thymine. Look at this molecular scenario and think about the possible problems in protein synthesis in this DNA mismatch. Suppose this protein destiny was either the heart or the occipital lobe, our vision region in the back of our brain; in all probability, it would provoke serious problems to our eye sight. Another area of great concerns for us is the pre-frontal cortex, the last brain region to attain maturity. It is vitally important in cognition, rational thinking- along with decision making. Most recently, researchers have found that the pre-frontal cortex is intrinsically connected with the limbic system in decision making. A rational decision is dependent on input from both brain regions. During adolescent hood, the time when we are neither an adult nor a child, hormone release is triggered by simple stimuli, emotions arisen from the limbic system. The emotional stress and physiological demands we go through during this phase of our life development and growth is, perhaps, the most critical during our entire life time. Stress may be provoked by many things inside and outside us. Unconsciously, we may be under continuous stress caused by childhood painful experiences that predisposes a person to multiple illnesses later in life. Stress may have come from parental disputes and illness, natural disasters like Earthquake, tsunamis, fires, poor nutrition and civil strife. A poor diet may contribute to inadequate consumption of amino acids rich food. Amino acids are the building blocks for proteins and the latter are our body building blocks.

MORE ON DNA METHYLATION

Before we proceed to elaborate another theme, we would like to say a few more things about methylation. DNA methylation is a modification that controls gene expression. This is a very important and strong statement. It says that controls gene expression. If it controls gene expression, consequently, the whole process of protein synthesis comes under its influence. Methylation, which is a molecule composed of one carbon atom and three hydrogen atoms bonded to DNA, seems to be a powerful molecule. Besides, DNA methylation contributes to mammalian development, aging and some diseases. In humans and in mice, our lovely research tools, the addition of a methyl group to cytosine within a cytosine- guanine dinucleotide is catalyzed by DNA methyl-transferase and enzymes DNMT3A, DNMT3B or DNMTl. The latter is the maintenance methylase because it adds a methyl group primarily to double strand DNA that is already methylated on one strand. I promise not to repeat it again, but the cytosine-guanine dinucleotide the researcher is referring to are the letters T and G that we talked about in the double helix. When a methyl group is added to DNA bases, it will modify the sequences, for example GAAT. The methyl molecule CH_3 attachment to the letters just mentioned will make the letters group (gene) look different from surrounding similar group of letters. Consequently, restriction enzymes that make a cut when they recognize a particular sequence will bypass the sequence GAAT. This sequence of letters became under the control of methylation. Methyl is not the only DNA marker, and they play multiple roles. Do not hesitate reading it, the next paragraphs are much easier. You must keep in mind that you have brave and extremely smart cells in your body. Your DNA molecule plays around with four letters or bases, amino acids, RNA and multiple molecular transformations before a protein comes out to form part of your eyes, heart, brain or legs for you to move around. All these are very complicated processes that are taking place inside you. You can say you are doing it because nobody else is around to do it for you.

Chronic stress and your brain

The pre-frontal cortex is often associated with schizophrenia, bipolar depression, anxiety and a host of other brain disorders. It seems the cell has insurmountable problems to conquer to be able to survive. Of course, it has developed its own repair system including demethylation. J.D. David Sweatt at the University of Alabama at Birmingham, doing research on methylation wrote that demethylation occurs rapidly under certain conditions, such as when people experience stress.[16] Chronic stress has been demonstrated to be extremely harmful to our mental and physical health. Brain organs like the amygdala on both temporal

---

[16]- Science News, 05/24/2008, Vol.173,#17, p.19

lobes and seat of learned fear is a brain group of cells prominently exposed to chronic stress.

The hippocampus on the right and left hemisphere of our brain and site par excellence for memory and learning is also seriously compromised under stress. Some researcher have found that chronic stress kills cells in both hippocampus, thus seriously compromising your ability for learning and memory storage and retrieval. An over-stressed amygdala would not be able to respond properly and fast enough when threatened from outside your body and, you may be an easy target for a predator. While in the jungle, the amygdala saved us from becoming the lunch or dinner for a hungry tiger, lion or being bitten by snakes. In big urban areas, the amygdala helps save us from human predators. Similarly, viruses and bacteria otherwise dormant, when your defense system is weakened by chronic stress, may seriously threaten your health. Most of us are aware of stressful situations and their harmful effects on blood pressure, heart problems as well as tension headaches. We usually run to the pharmacy to get over the counter medication to get rid of pain and discomfort from chronic stress, but we rarely stop to think about the harm we are doing to our body. Your entire immune system is seriously compromised under stress, particularly, chronic stress, and symptoms and diseases otherwise dormant under the watchful eye of T and B cells and an army of assistant cells. Bacteria and viruse diseases do not miss a chance to attack healthy cells and get us ill. One more note on the amygdala. In recent years, Science News, 02/26/2011, the surge of new research has expanded scientist's view of the amygdala's importance. It turns out the amygdala helps shape behavior in response to all sorts of stimuli, bad and good. It plays a role not only in aversion to fright, but also in pursuit of pleasure. It has neural connections with all five senses. This is a very primitive organ of your brain which assigns value to reward and adjusts that value as circumstances changes. It also suggests that both amygdala play a role in goal oriented behavior. Having connections with all senses, it is in a unique position to take immediate precautionary action when it is absolutely necessary. In addition, it will involve our commander in chief region of our brain, the pre-frontal cortex, for rational decisions. We must add that connections between the amygdala and the thalamus are very,

very close. If necessary, the axis composed of thalamus, hypothalamus, pituitary and adrenal gland will be pumping out hormones all over your body and my body. I am not aware that I am doing it, but my heritable genes that I got from my ancestors equipped me with emergency tools for protection against unforeseeable threats. You can rest assured that methylation is implicated during these processes.

Our body energy and wisdom

Your body and my body have in storage vast amounts of energy and a treasure of accumulated knowledge and wisdom. Have you ever thought of the amount of energy stored in each atom of your body? Ever since we dropped the first atomic bomb in Japan I have been wondering about the potential power of atoms in my body. Einstein, Bohr, Oppenheimer and the likes have put ideas in my brain that make me wonder how I exist on planet Earth. I am made of atoms that form molecules that will become cells and ultimately organs and tissues forming my body. However, the space in atoms in my body is more than 99 percent empty with electrons traveling in orbital fashion at nearly light speed. The subatomic particles traveling at such speed inside my atoms inside my body seems to me to be a walking powerful energy warehouse. I have very little control of what is going on in this chemical and atomic machine of mine. And… you are right, how can I say that this machine is mine? My organism is continually processing external perceptions and internal chemical and atomic reactions. When I am witness to a car accident on the highway and I see people injured, especially, if the injured are children and women, I will have millions of chemical reactions taking place inside me that I am not aware of. Sadness and fear might be the first feeling registered in my brain. The thought that I could have been involved in the accident places my amygdala in a state of alert. There is an amygdala in each temporal lobe. The amygdala will automatically alert my sympathetic nervous system for an immediate reaction if it is called for. If it is life threatening, the amygdala will either fight or flight if it does not chicken out and freezes. In the mean-time, the amygdala will connect with the thalamus that in turn will get in touch with the hypothalamus which will release a message to the pituitary gland which will release a hormone down to the adrenaline gland. The adrenaline coming from this gland will move up the body and trigger a myriad of responses through millions of chemical and electric reactions. Needless to say, my limbic system and pre-frontal cortex must have been involved in all decisions in my brain, except, perhaps, the split second reaction of the amygdala in a life threatening situation. The point that I am trying to convey through this hypothetical illustration is that a visual stimuli with no matter attached to it, is capable of provoking in my organism (maybe in yours too) a series of behavioral responses in which matter and electricity were involved. Matter was converted into energy, and energy, like in neurons, is converted into matter. I do not know an iota how this transformation takes place, but rest assured, it is taking place every fraction of a second in my body. This stressful situation I just described for you, compromises my health, among other things, the immune system. My heart palpitations went up, blood pressure, asthma symptoms and a headache became an unavoidable health threat. You may dismiss this health lesson as a trivial every day episode of modern life style. At issue is your own reaction to daily life episodes. How many stressors do we encounter in a day, a week, a month or during a year? In this biochemical scenario, your DNA molecule, RNA and its many micro molecules are modifying the structure and function of cells, organs and tissues in your body. A hyperactive amygdala under chronic stress can be a serious health problem. An excess or deficit of a neurotransmitter in the brain can lead to a brain disease. The sooner we become aware of genes, epigenetics and environmental causal factors, the better for our health and a happy life.

Through worldwide network studies, researchers are sharing genome connected information regarding diseases. It allows a researcher or research teams to follow up on another person's findings and ideas. Not all researchers have access to money and technology to go to the next level of investigation in their project. Some countries and institutions provide research funds only when the by-product is close at hand and has an immediate and necessary use, like in disease and food production for its population's needs. Long term basic research, if not ignored, is pushed aside. In the U.S.A. and Europe, basic research takes in a good sum of money. The sequencing of the human genome, restriction enzymes, recombinant DNA, gene therapy, deep brain stimulation, just to name what pops-up in my brain, are all American accomplishments. James D. Watson put the double helix together in England, but he is an American by birth. Most recently, our scientists are exploring how to engineer genes to make proteins at the moment and location we need to have them. A new research tool dubbed ontogenetic would allow neuroscientists not just observe RNA molecules respond to DNA messages, but will empower it to manipulate neurons at will. Gero Miessenbock at Oxford University in England says: Light responsive molecules used in ontogenetic experiments uses a specific wavelength of light... a channel opens and allows positively charged ions to flow into the cell. This happens to be the neural code for <u>on</u>, K Deisseroth from Stanford University said. Other light responsive molecules, when tickled with the corresponding wavelength of light will allow negatively charged ions into the cell. Using a combination of the two types of molecules and different wavelength of light, researchers can flip neurons <u>on</u> and <u>off</u> at will to find out how neurons interact with their neighbors.[17]

Hopefully, this new research tool will allow us to look at neurons in the hippocampus and actually observe the neuronal behavior in learning and memory formation. Similarly, focusing on the amygdala, it will show you learned fear. The nucleus accumbens and the tegmental ventral area in and around the midbrain will teach us how addiction behavior is interconnected in that area of the brain. We are moving forward at great length after the discovery of the double helix and the genetic code. We are in pursuit of our goal deciphering and making corrections on gene misspelling, mutations as well as translation processes multiple mistakes.

With your permission, we are going back to SNPs and missing nucleic acids letters like A T C and G on DNA as it pops-

---

[17]- Science News, 01/30.2010, Vol. 177, # 3, p.18-21.

up very often in research literature. It is well known that tagging-binding- a SNP such as a methyl group to DNA, generally suppress, meaning shuts down, genes next to it. We already mentioned the possible harmful effects of a missing or altered genes in protein synthesis. Rare mutations- single letters DNA changes- can contribute in a big way whether a person gets a disease or remains relatively healthy. Some variants have been connected to common disease like schizophrenia, bipolar depression and mental retardation. Determining the genetic base for any of these diseases will take time and perseverance. It is not only a matter of genetics, epigenetics and nurture, but the conformation of the disease itself. In schizophrenia you may rightfully argue that the whole brain is involved in producing multiple symptoms. It is cognitive, affective, visual and auditory, besides covering smell, taste, skin and other symptoms that are hard to attribute it to a single word of misspelling in the DNA alphabet. However, general wide studies with many participating researchers engaging thousands of patients worldwide, might point the right direction while pursuing a particular disease symptom pathway.

We love to repeat teaching lessons from James D. Watson. He did not stop working for us. His goal was not to leave for us the double helix, but the sequencing of our genetic code as it relates to our health. He wrote: A single base change in the DNA sequence of human beta hemoglobin gene results in the incorporation of the amino acid valine rather than glutamine acid into the protein. This simple difference causes sickle cell disease, in which red blood cells become distorted into a characteristic sickle shape.[18] Watson was 'mind traveling' in space and time during his work at Cavendish Laboratory along with F. Crick in the early 1950s. As I was growing up, I remember we used a sickle shape blade to clear weeds and small bushes growing under coffee trees. In some Latin American countries, the economy is based on sugar cane, tobacco and coffee crops. Therefore, the sickle shape blade was a valuable working tool.

---

[18]- James D. Watson, op cit., 67.

Michael's inquisitive brain

Once more we will repeat this mutation process to satisfy our inquisitor, Michael's brain. We need three nucleotides or letters of DNA alphabet which normally has four: A.T.G. and C. for binding to each other. Three letters specify to form a codon. In most cases the genetic code for each amino acid can be multiple codons composed of three different letters. For example, the amino acid leucine can be formed by six different codons while tryptophane is formed by only one codon, namely U.G.G. Scientist Vernon Ingram at Cavendish Lab. had found out that amino acid glutamine formed by codons G.A.A. or G.A.G. found at position 6 in the normal protein chain, was replaced by valine that can come in four different codons of the genetic code: G.U.C., G.U.U., G.U.A., and G.U.G. in sickle cell hemoglobin. You must remember the four letters found in DNA and its difference to the RNA chain. RNA has uracil as a nucleotide, while in DNA thymine is a complementary nucleotide to adenine. In RNA, it is uracil that complements adenine. You have these two great molecules working together to build tissues for your body. You have DNA polymerase that unzips DNA during the replication process; actually, it is DNA helicase which unzips it. However, during the transcription process in the making or forming of a messenger RNA, the RNA polymerase helps creating a single strand. Messenger RNA carries the message, and is the template for protein synthesis.

This messenger RNA that was transcribed from the DNA cell nucleus will end up at a ribosome. The sequence of the messenger RNA will be used at each ribosome to form or synthesize a new protein molecule. Another molecule, transference RNA, carrying amino acids will cross the cytoplasm and anchor itself at ribosomes. At one end of this t RNA there is a set of letters, let us say, U.U.C. that recognizes and establishes a bond with its opposite codon: A. A. G. in the messenger RNA, with the resulting codification for lysine. An RNA molecule serves as a template for translation of genes into proteins. Another RNA molecule transfers amino acids to ribosomes, the assembly site for protein synthesis; besides, it conforms into other useful small forms. You have also seen that multiple codons can specify for the same amino acid. This dynamic process could create problems in protein synthesis. Take, for instance, leucine with six different codons and tryptohane with only one codon. It would appear that protein folding from tryptophane only one codon would be much less mistake prone than six codons, leucine. Now consider variants like nucleotide polymers components, its availability in the cell, and the intervening stages in codon selection, availability, peptide chains and final protein folding, it will provide enough room for multiple genome mistakes and consequent diseases. You may want to accuse us of picking on translators, but this RNA transfer molecule has to do a superior job not to commit mistakes culminating in dysfunctional protein and diseases. Improperly folded codons, competing codons, interference RNA and micro RNA, just to name few of the possible road blocks in our endeavor, are presenting possible risk areas in any mammal genome sequence study. We hope to have given you a good idea on how this game of nucleic acids letters are playing around forming proteins for your body while you are oblivious of complex processes taking place in your own body. A couple of pages back we mentioned to you how this transportation system, seemingly so easy, can be threatened by simple small nucleotide polyphormisms (SNPs) and interference RNA, preventing genes from carrying out their job of protein synthesis. These SNPs molecules love to mess around with the thymine nucleotide in the double helix and create unwanted problems for us.

EPIGENETICS

Epigenetics, the nucleotide process that may control a gene potential to express itself has come to take front page in some journals because it has been implicated in some common, but hard to break diseases. The most common and best studied form of epigenetic intervention in a DNA molecule is methylation. It has the potential to manipulate the cell to ignore any gene in a stretch of DNA. We are subject to epigenetic interference from very early in our life processes according to some researchers. It can begin during our embryonic phase of development. During this early stage of cell undifferentiation, let us say, blastocyst, stem cells accumulate, among others, methyl groups and direct stem cells into one of the three germs layers. Each of these germs layers produces a different set of adult tissues. These tissues we are referring to may be protein for your heart, lungs, eyes or brain. Making epigenetic a little more interesting, but headache provoking, a group of researchers at the University of Sydney in Australia discovered that methylation of the fur color genes (in mice) persists in the female germ line, allowing it to be passed down to off-spring like a change in DNA.[19] Local and worldwide studies share their common interest in the human genome, epigenetic, and particularly, methylation interference in DNA expression. A lesson to be seriously considered during disease research etiology is the role SNPs play in all its forms, including methylation, interference RNA, and micro RNA during embryonic and fetus development. The cell has proven to be an excellent self-sustained unit of life, but there are many internal and external factors that threaten good health. Whether it is endogenous or exogenous, the great DNA molecule has survived millions of years. It developed brain cells for intelligence and critical thinking to assist you get rid of temporal nuisance. This brain is developing very sophisticated tools for self-correction and to accelerate its own growth. Among the beauties of our recent technologies to improve our health is the decoding of the human and mouse genome a decade ago. Mouse and human have about the same organs, although they separated from each other some 70 million years ago. Most scientists agree that a mouse genome sequence is about 95 percent that of a human. We cannot play around with a human as a research scientific object, but a mouse can be carried with you even when you go on vacation. Gene-targeting technology using a strain of mouse, a mouse we can knock-out a gene in any organ of the mouse and peacefully and intelligently, observe causal effects without any pain to a human being. Besides, a mouse does not demand dormitory, dining and recreational facilities. The researcher and his/her expensive tools is the expensive subject. These knock-out mice are very important in medical research. A mouse as an object for gene studies is relatively new. The first reporting of this technology stating that it could generate gene-targeted mice, was

---

[19] - Scientific American, 8/20/2010, Vol. 303, #2, p. 2224.

published in 1989. Twenty one years later, three pioneer scientists working independently of each other were awarded the Noble Prize for Medicine or Physiology. The New York Times, Science Section, 10/09/2007, broke the news for us. The three pioneer scientists are: Martin Evans from Cardiff University, Wales; Mario Capecchi from the University of Utah and Oliver Smithies from the University of North Carolina.

Sequencing genome from different species can tell us how each one of us is different from each other's genotype and phenotype. It can point out mutations and diseases, but, perhaps most importantly by comparing genomes from different animals, can tell us gene and gene segment that have remained unchanged for many years. Many years may be counted as meaning millions of years. We can date animal separation from each other based on a set of genes that have remained un-altered over millions of years. Observing gene locations for protein expression placed side by side taken from different animals, we may be able to tell how a species changed into a different phenotype. However, for us now, the immediate benefit of comparing genome sequences is the increased precision with which researchers can reveal the sequences that have been carefully preserved over time, implying that they have an important role in the organism. Alternatively, these comparisons can pinpoint sequences that differ in just one species or group of species.

More on epigenetics

To some of you we may sound unnecessarily repetitious and perhaps, cognitively naïve. Our apologies to you, but our classroom experience and supervision taught us that going over a subject or a lesson is not a waste of time, but the strengthening of synapses. We will define or describe epigenetics as the study of heritable changes in gene expression that are not the by-product of changes in DNA sequences. We briefly addressed epigenetics indirectly when we mentioned SNPs and gene modifications. We also spoke about the chromatin complex composed of DNA and histones with its little tails extending out as possible SNP sites. We attempted to connect it with current concepts of on and off protein switches. We could easily transfer this concept to our electrical system of on and off switches. It would sound and look neat, clean and easy to follow, however, we are dealing with living acids like DNA and RNA capable of forming multiple tissues and organs. During these processes, including before and after transcription, many modifications take place altering the ultimate by-product, protein synthesis. Where and how these modification take place has been the subject of curiosity and researchers during the last decade. Small nucleotide polyformism molecules attach to chromatin, most often, linked to or hanging on to histones proteins, are called epigenetic marks by most researchers. Many unanswered questions are still lingering around the concept of the on and off enzymes switches. Are epigenetic modifications passed on down to off-spring through cell division? What about modifications that take place during the transcription phase where the two great acids, DNA and RNA are involved? Before we proceed to create more questions, which are not difficult to make, let us deal with chromatin modifications. One of the most frequently used concepts to explain: On and off switches, is that a DNA and histones modifications are consonant with either negative or positive transcription states. In histones post transcription modifications, cytoplasmic response to transcription factors- positive acting marks- are established during the gene activation. In histones post translational modifications, it is in response to cytoplasmic signaling to transcription factors that positive acting marks are established. It takes place during gene activation by recruiting relevant enzymes by DNA bound activators and RNA polymerase. Similarly, negative acting marks are established across genes during repression by DNA bound repressor recruitment.

This seems to be a simple and attractive model to conduct chromatin research. However, some researchers claim that it is not as simple as it appears to be. The technology of on and off enzyme switches is not a matter of connecting and disconnecting cables or just turning on genes. Many of the marks we have been talking about play multiple roles. It is a dynamic metabolic process in intracellular chemical reactions that converts an organism's infinite number of molecules into energy producing action units.

Chromatin marks are more complex molecules than anticipated. It responds to and performs multiple complex and different roles. It is now a belief among some researchers that a chromatin's role in transcription is a mechanism in which DNA and histones modifications establish gene activation and then re-instates repression. Thus, DNA methylation and histone modification is a very dynamic process, it precludes a fixed status. Some marks now seem to recruit both, activating and repressing effector proteins. Besides, positive and negative acting marks are established during transcription. Two types of chromatin modifications for the regulation of transcription of the protein- encoding genome are activation and repression. These DNA modifications seem to take place at specific junctions of nucleotides and at different locations.

Epigenetics and post translation modifications

There are a good number of post translation modifications. Most take place in the amino and carboxyl terminal histone's tail. Among common modification sites, Lysine is a key substrate residue in histone biochemistry. It undergoes many exclusive modifications. This really seems to be the Achilles tendon we were talking about at the beginning of this essay. "In general, the functional consequence of histone's post-translational modifications can be direct, provoking structural changes to the chromatin. Changes can take place by recruiting proteins."[20] Epigenetic changes are decisive, indispensable during the development and differentiation of different types of cells in any organism and normal cellular processes. However, like in all system, an inappropriate epigenetic modification can lead to serious health problems. Epigenetic processes may be involved in very sensitive areas as in a cell shape and function. It may intervene in chemical modification to the DNA and histones in the chromatin packaging. This macro-molecule-DNA- has taken appropriate measures to protect itself from harmful intruders- its job is awesome. The protein (histone) that DNA wraps around forming the chromatin structure, in itself has gone through multiple processes. Proteins are the most abundant molecules in our body.

---

[20] - Nature, 05/24/09, Vol. 447, #7143, p. 407-412, by S.L. Berger.

The number of proteins in our body makes it the most likely target for changes, modifications and assault by internal and external forces.

'Epi' means over, or in addition to; genetics, is now days used to mean the study of inherited- heritable- changes in our appearance. In genetics jargon, it is known as phenotype. The change may also appear in gene expression, but not involving changes in DNA sequence. At present time, the word heritable is paramount in epigenetics as long as it does not involve changes intrinsic, meaning basic, to DNA sequence. You may rightfully ask, how can it be heritable, if it does not involve DNA sequence?

Well, epigenetics has become a field of study of biology in itself. Epigenetics like RNA's multiple forms and functions has attracted brilliant biology researchers hoping to find and characterize toxic proteins that threatens our health. We have just finished sequencing the human genome; searching for tools to destroy pathogens that threaten our organism is our next challenge. The job ahead is worth our interest and enthusiasm. It took our most primitive cell millions of years to evolve from a simple cell to a complex organism like yours and mine. Among your body's multiple organs, you have a brilliant brain that is engaged in finding solutions to diseases that threaten your life. Your brain, when not engaged in fighting diseases, amuses itself trying to become a painter like Picasso, a violinist or an astronaut navigating outer space.

Maternal nurturing and DNA methylation

Our main concern during this essay has been centered on how the genome sequencing can improve our health. There have been well documented studies in mammals reporting that maternal nurturing alters DNA methylation at the gene encoding glucocorticoid acid receptor. In the absence of appropriate nurturing, the chemistry of DNA in her offspring is seriously compromised or placed in jeopardy. If normal DNA methylation is negatively altered in genes coding for proteins destined for the hippocampus, the child's learning ability will be placed in jeopardy. The same will be true for negative DNA methylation for

45

genes coding proteins for the amygdala. In all probabilities, the child will not be alert enough to appropriately respond to environmental learning signals. Whether it is the hippocampus or amygdala deficient functioning secondary to an abnormal DNA methylation, the child's general health would be in great risk. Applying the definition of epigenetics as heritable changes, although not involving DNA sequence, helps us understand and deal with learning difficulties and emotional problems in our schools and general population. A considerable number of research studies have raised the profile of epigenetics making it a particular field of study in biology. Hopefully, it will help us understand and provide appropriate treatment for diseases like Parkinson, Alzheimer, schizophrenia, autism and violent behavior leading to bloodshed and fatal episodes that recently have taken innocent life in our society.

Cancer and epigenetics

A report from Epi Gen Western Research Group at the University of Western Ontario dated 01/31/2006 defined epigenetic as the study of heritable changes in gene expression that occur without a change in DNA sequence. By now -2011- this definition has been established worldwide as the gold standard for research and findings on epigenetics. The leading author of this report, Dr. David Rodenhiser states that "epigenetic mechanisms are critical components in the normal development and growth of cells. Epigenetic abnormalities have been found to be causative factors in cancer, genetic disorders and pediatric syndromes as well as contributing factors in autoimmune diseases and aging." This quotation places epigenetic abnormalities as a causative factor for cancer. Cancer has been under the microscopic eye for quite a long time. However, there are different types of cancer appearing in different tissues and organs of the body. In man, we have prostate cancer that tends to stay localized, while in women, we have breast cancer that often metastasizes. We also have colon cancer that can be removed by surgery. We have throat and lung cancer widely associated with excess smoking. There is cancer whose etiology is attributed to environmental factors like toxic gases, air pollution, water contamination, pesticides, and even chronic stress has joined the list of cancer risk factors. However, most these cancers just named are of a temporal nature or etiology and do not classify in the definition of epigenetics because it is not passed on to offspring. The key word, heritable, cannot be applied to a cancer in a person with a long history of smoking a package of cigarettes a day. The same would be true of a person exposed to toxic or radioactive chemicals at his place of employment. There has been enough evidence to connect lung and throat cancer with cigarette smoking. The leading author in this article proceeds with DNA methylation and histone modifications describing the basic components of a nucleosome, namely DNA wrapping around clusters octomerous (from the Greek okto=8 and omers=parts) of proteins-histones.

Genetic markers

This coiling of DNA around histones is widely known as the chromatin structure. Changes to this chromatin structure influences gene expression, the gene expression we are referring to is protein synthesis. You are aware that if protein synthesis is jeopardized, it would be serious trouble for you and I. Protein shape determines function and the if the shape of a protein whose final destination is the heart is altered, specifically, a valve in the heart, if the fetus survives in the mother's womb, as an adult, he or she will develop heart problems. We should add that cells have their own intrinsic repair system, but unfortunately, the repairing system does not work to our advantage every time we are in need. You can also take a protein destined to be part of an axon in a brain cell. Axons are engaged in electro-genesis; its functional components are sodium and potassium pumps that will send a chemical message across a synapse to a nearby post-synaptic cell membrane receptor. Any process or epigenetic mark that messes up the formation of proteins is threatening my life and health. Besides, you are already acquainted with the negative resulting effect of missing genes, segments of genes, the change of one nucleic acid for another and more; they all provoke health problems. Further elucidating genetic marks, the author named some of the enzymes involved in the process such as: DNA transferase, histone deacetylase, histone acetylases, histone methyltransferase, and the methyl binding domain protein, (the ending -ase denotes an enzyme). Dr. Rodehiser alerts us that alterations in these epigenetic patterns can deregulate patterns of genes expression, which results in profound and diverse clinical outcome. Some research finding in David Rodenhiser's article called our attention because it has been confirmed by other research teams. DNA methylation involves the addition of a methyl group, CH_3, to cytosine within CpG. These last three letters means cytosine/guanine pairs. Cytosine is a nucleotide to both molecules, DNA and RNA. The point we are trying to show you is the importance of this epigenetic mark involvement is in critical genes processes. Another emphasized point I would like you to remember is that changes in DNA methylation may occur as a result of low dietary levels of folate, (a B vitamin), methionine, which is an essential amino acid, or selenium, which can have profound clinical consequences such as neural tube defects, cancer and atherosclerosis. This last observation was supported by five different researcher groups. We hope we do not sound excessively redundant, but among the highly suspected contributing factors altering DNA methylation and histone modifications contributing to the above mentioned health risk are air pollution, water contamination, pesticides, chronic stress, excessive exposure to dangerous chemicals and psychological causal factors. A note of reference on our epigenetic readings, we want to alert you in advance, is that most of our quotations on epigenetics are dated up to 2006. A lot of research has been done and published during the last five years we hope to cover. When the double helix structure was made public, who would have believed that at present time we can read the structure and instruction to put together a human being; make corrections on genetic diseases and create a synthetic bacterium? We are close to making science fiction a reality by creating artificial intelligence. Even splitting an atom was not only considered impossibility; but a threat to all forms of life on planet Earth.

The DNA structure in the mind of our scientists

The idea of the DNA structure and function occupied the minds of a group of brilliant young scientists right after WWII including our star chemist, Linus Pauling. However, the Britons Rosalind Franklin, Maurice Wilkins, Francis Crick and the American J. D. Watson were determined to give the United Kingdom the prize of stardom in molecular biology. Although Rosalind seems to have been ahead of the group in designing the double helix for the DNA molecule, it was Watson who did not miss out a single opportunity to check out everyone else's ideas and try to put it to work. He was checking out every article in biology and related sciences that might give him the final answer. He argued alone how the spiral strands should go, inside or outside, with its nucleic acids hanging on like necklace beads. On the other hand, Rosalind may have been closer than anybody else to the DNA double helix structure, but she insisted on more experiments before announcing it publicly. Watson's disposition to share his curiosity and observation with Crick did not miss an opportunity to pick up cues from other researchers that might have advanced his own insights. Rosalind's crystallographic pictures struck him like no other picture had. He finally figured it out that it must be a two strands structure with nucleic acids beads inside chemically attracting each other. His approving alter-ego, F. Crick, gave him his blessing and the structure of our genome soon was going to take front page on the most prestigious science journals, newspapers and television networks. I, for one, would have thought that Watson and Crick were celebrities to be paraded on Broadway in New York City and in Washington D.C. Well, things do not always come to pass the way you like it to be. Even his Nobel Prize did not come until 1962. As if it helps anything, the person behind all these genetics, Gregor J. Mendel, a monk was not recognized even by his own colleagues in the monastery. In Watson's case, researchers working on and around his discovery were awarded the Nobel Prize for subsequent discoveries relevant to DNA structure and function. Watson and Crick had to wait for another few more years to be called in from the Nobel awarding Committee.

What was so mysterious about this double helix structure? The solution of the mystery would unlock the key to the creation of life. They proposed that each nucleotide, namely, adenine, would chemically attract the opposite complementary strand nucleotide, thymine. Bingo, this is it. Now they needed to proceed with protein synthesis. How are these four letters, A.T.C.G. of the double helix, arrange themselves to make the 20 amino acids necessary in the human body into proteins? Crick began to work with Sydney Brenner and together they demonstrated that the code was based on a series of three bases named codons. Researcher Marshal Nirenberg and John Mathei identified the first letter in the genetic alphabet. They reported that the amino acid phenylalanine had the code, UUU. Please, turn to the table at the end of the book showing amino acids and their codons code. You can observe that most amino acids have more than one codon or triplets codes. In addition to the 61 codons for amino acids, there are three codons for a stop command. The stop codon indicates the end of an amino acid sequence coding for a given protein. Most proteins are comprised of 100 or more amino acids, and so a long sequence of codons is needed for any given protein."[21]

Watson and Crick parade in the Heroes Canyon

After discovering all the codons for each amino acid that will form peptide chains, which in turn, become proteins for cells and tissues in our body, what was next? In other words, it is a sequence of codons that codifies the production of a particular protein, which in reverse definition, it is a gene. These sequences of codons that codes for an enzyme will be regulating neurotransmitters in our brain. A neurotransmitter, like dopamine, which is involved in very serious brain diseases like Parkinson and Alzheimer, is also engaged in making us "feel good." This brain chemical is often associated with feeling "high" and addiction. This discovery opened up not only a new field of research-molecular biology-but another approach to therapeutic medicine. According to Crick's Dogma: DNA-RNA-PROTEIN, it is the DNA nucleus that will be issuing orders for proteins synthesis in the cell. These proteins are the building blocks of our body. No

[21]- Brave New Brain, Nancy C. Andreasen, Oxford University Press, 2001, p.101.

wonder I was expecting a big parade for Watson and Crick through the Heroes Canyon on Broadway in New York City.

In retrospect, we thought that as soon as we had learned the step of codon sequence and protein synthesis, we could begin engineering or coax cell organelles and construct a miniature life system. I was not too far off; C. Venter in the year 2010, announced publicly that he had produced a synthetic bacterium. And…James Watson was around to celebrate it. Venter demonstrated that we have mastered the mechanism of DNA-RNA molecule secret to make live cells that will grow to become a whole beautiful organism like Dolly, the cloned sheep. Perhaps we will have to wait for Venter's disciples anywhere in the world to see the application of his newly discovered tool used for therapeutic medicine and pharmacology. We love being redundant. It is that in just a little over a half a century we have gone from mule riding to supersonic air ships. When I was attending elementary school, a trip to the moon was thought of as riding on a broom stick. It was a witch story associated with evil spirits.

The very old RNA and DNA molecules

The DNA molecule began life many, many years ago either by chance or just basic chemical and functional need to pair with or incorporate another great molecule, RNA. That union facilitated, among other things, the formation of a protein and cell growth into a multiple cell organism. Definitely, it was a giant step forward in evolution. Equally important was the annexation of another important molecule, the mitochondria. When and where these wonder molecules fused together, is basically an intellectual mental exercise. DNA survives by self-replication or duplication which is the same thing. During replication, it divides in two strands which are self-template for transferring the entire database into the new cells that will create new organisms. Second, DNA functions take place in the cell nucleus. A single DNA strand-messenger RNA is formed which eventually will carry the message for protein synthesis at a ribosome. Another RNA strand named transfer RNA will be traveling to a ribosome. Out of this assembly station, the ribosome- peptide chains will come out forming proteins. This linear chain of peptides will be shaped into proteins for different organs and tissues in our body. There are many proteins performing many different functions in different parts of our body.

Introns and exons

Within the genome, we have coding genes and non-coding genes. The old name for non-coding genes was junk DNA. The new name for non-coding DNA is dubbed introns. In contrast, exons have the codes that determine the sequences of amino acids that will provide the final protein. This is not an easy and simple job. Just imagine the complexity of a template for a protein destined for my left foot toe and a template for a protein in the pre-frontal cortex. As already mentioned on above pages, most of our genome consists of introns, or as you may say, junk DNA. Some researchers claim that over 90 percent of our genome is non-coding material. An interesting mechanism exists within each cell. During the transcription process which produces the mold for protein synthesis, meaning the messenger RNA molecule, introns are thrown out. They are removed before they are transported to a ribosome where proteins are made. Introns do not have a job in the translation process. They are non-coding genes. Researchers called introns removal: splicing introns out before they can mess up our protein assembly factory.

The transfer RNA strand carries three nucleotide sequence known as anticodon. It facilitates the matching of m-RNA and t-RNA at ribosomes sites during the making of a protein. Just stop for a few seconds and think about all the different proteins in our body. Proteins responsible for making up our brain cells have to, among many other functions, generate electricity to produce impulses that will secrete neurotransmitters that, in turn, will activate another cell.

Your memories from birth to death will be stored in proteins. And…how marvelous, the instructions or command from all these proteins, each one with a particular function, comes from the nucleus of the cell. This great molecule with its powerful nucleus is a very old one and has proven to be immortal. It has been feeding on chemical elements existing on planet Earth for billions of years. Our body is a huge and complex chemical machine or factory that has survived Earth-quakes, tsunamis, hurricanes, electrical discharges of immense power, multiple iced ages and a huge meteorite about 68 million years ago. This cell is working from conception until death. It rests while we sleep although our vital organs are on a 24/7 schedule, while genes are turned on and off as needed. Sleep is not an accurate term for us to deal with cell activity. For instance, take the hibernation cycle. Some small animals hibernate for months or even years, deep in the ground. This is an excellent research theme for sleep and insomnia studies. Somatic as well as brain cells are close to zero activation, but still alive. Buried deep in the ground they are not responding to outside stimuli like wind, rain, sunlight, drought, but maintained by a balance of inside Earth temperature and whatever is going on above it. Cells are at a very minimum of activity to be able to survive when the appropriate time comes. Are they sleeping, are they unconscious, or in a state of comma, hypnotized or just a bundle of atoms clustered in molecules that will begin chemical attraction with a little touch of nature? Here we may find the answer to the beginning of life on Earth.

THALASSEMIA, HEMATOPOIESIS AND RELATED CASES

Genes may be turned on to produce protein to consolidate a memory permanently or to reconstruct a broken tissue-ligament- in my left toe. Genes may be called in to produce enzymes that will provide chemical reactions that in turn will trigger serotonin release and make me feel sad. All these switches —on and off switches- proteins may be provoked by internal or external causative factors. I have seen a puppy ran over by a car trying to move while dragging its paralyzed legs behind him. I caught myself wiping tears from my eyes. My body was responding to a little dog in pain that I had never seen before. A visual stimulus had triggered a very complex response process I could not control at a conscious level. The cell nucleus was responding to, and commanding multiple actions that I was not aware of. Besides shedding tears for a little dog I had not seen before, I inherited thalassemia traits from my parents. I passed these traits on to my daughter. We were eleven children; five inherited it, one died before he reached three years old. In thalassemia, there are not enough red blood cells to carry oxygen. The protein hemoglobin that transports oxygen from the lungs to the tissues is made in the bone marrow by stem cells. When my body needs to make more of the protein hemoglobin, the pertinent or applicable part of the bone marrow DNA – meaning- the hemoglobin gene opens up, or as the professionals say, unzips. During this unzipping process, only one strand is copied. This process is also known as transcription. Here, a protein called RNA polymerase lends a hand in the process and, the single strand coming out of the bone marrow DNA is the single strand the original biologists called messenger RNA. This is the template for the production of a protein. We hope that in a near future my daughter and family members would benefit from gene therapy and eliminate that health risk from our protein factory. Protein is put together during the translation process. The command to produce protein ultimately comes from the nucleus. However, what or who is responsible for alerting the bone marrow DNA that there is a need for more hemoglobin, this author does not know yet. We do know that proteins turn genes on and off at certain locations. We would like to quote from the New York Academy of Science 2005 report on hematopoietic stem cell research. "Despite the intense research, many long standing questions of experimental hematology remains unsolved…Real time tracking of individual cells in culture, tissues or whole organism would be an extremely powerful approach to fully understand the developmental complexity of hematopoiesis. A vast array of cellular and molecular tools, RNA and protein expression data, and experimental methodologies have been developed for it and are already available for application. However, despite its therapeutic importance and long standing research efforts, hematopoiesis research has been unable to satisfactorily answer many questions"[22].

---

[22] -New York Academy of Science, Vol. 1044, 2005, p. 201--02

We have kept our interest on how genes are turned on and off, meaning how genes are called to produce, in my case, hemoglobin proteins for oxygen transportation. We reasoned out that genes respond to internal and/or external triggers or stimuli to produce a protein. In the process of protein synthesis, you must have an amino acid group at one end of the transference RNA. We looked for an amino acid in vitamins, minerals and dietary supplements rich in amino acids. However, our greatest concern was how to turn on the hemoglobin producing gene. At this level of research, we concluded that the concept of an on and off switch was very simple concept to rely upon. The whole process of protein synthesis involving many functional organelles within the cell, points to complex system of cell metabolism and chemical intervention. While I was watching my son Joe work on an electrical switch in the basement of his home, I asked him how he would relate the electrical switch he was working on with a gene switch in our cells. He directed me to the Institute Pasteur in Paris. He added that while he was studying medicine in Spain, a friend of him, named, Alejandro, had gone to Paris for a short trip. His friend had good connections and was able to visit the Pasteur Institute. Working on a different type of switch was a couple of brilliant scientists, Francois Jacob and Jack Monod. Our intestinal parasite, bacterium Escherichia Coli, E Coli for short, was receiving their undivided and focused attention. They found a repressor (a protein) that by binding to E Coli DNA, the repressor protein prevented the enzyme that makes up, or put together messenger RNA, from doing its job. In guerrilla warfare, it means that this interference during normal protein processing, was a sabotage episode. The process of transcription had been interrupted by a molecule at a DNA switch. This was eye opening, if not brain boosting. Protein shape determines function. We concluded that a DNA child, meaning a protein, could also play a role in regulating gene activity. This was another avenue of study, perhaps an RNA tool to engineer or manipulate our own genes. Monod and Jacob's discovery made us even more curious, if not restless, about the behavior and function of this RNA molecule. It is playing many undisclosed games, right inside our body. These two scientists were working on my intestinal tenant, E Coli. I was very happy to learn that soon, we may have good news on how to engineer it at will. How E. Coli lives in my intestines without paying rent besides feeding on my food, I do not know. E Coli has become an excellent research model worldwide. If we provide E Coli food and shelter, it justifies putting it to work for us.

However, the ribonucleic acid was interfering with the building blocks of my body. For example, transference RNA can block protein synthesis, micro RNA can do likewise and much more. From the very beginning I became suspicious of RNA. Too much power in one person or object gives me goose bumps. Well, there was one person we could turn to, Francis Crick. Most researchers call him the father of the dogma, DNA-RNA- protein. Is it DNA that makes use of RNA to do its job, or is it the other way around? How would you answer this question? If you have any doubt, even peptide bond links --one amino acid linked to another-- is done by RNA. We rightfully respect DNA cell nucleus to send out messages and command orders for protein synthesis. However, the fact is that there is an RNA molecule that can hijack the messenger strand and prevent the normal formation of a protein in my body is worrisome for me. It was more than I could swallow at the time. Anyhow, there are enzymes, again, produced on orders from the cell nucleus, which can influence gene expression. In fact, enzymes play significant roles in a cell metabolic and chemical reaction. In order to bring to closure the DNA- RNA dispute, we are going to refer you to a very reliable source that says, "The reality of the RNA World, an RNA dominated stage in the early evolution of life prior to the evolution of coded protein synthesis, has been firmly established by recent studies on the ribosome.[23]

Harry Noller and RNA

A scientist who could not resist the temptation of further studying the function of RNA, Harry Noller, stripped the protein

---

[23]- The RNA World, by R.F Gesteland, Cold Spring Harbor Lab. Press, 3rd edition, New York, p. 57

assembly station, the ribosome, of all its protein. However, despite lacking the protein that is normally part of its structure, the ribosome was capable of forming peptide bonds. Returning to Jacob and Monod's laboratory and their prized research model, the bacterium that made our intestine their permanent place of residence, E. Coli, they came to the conclusion that DNA must have a control system. Actually, the control system is a protein to activate a gene, which will say: "lets get to work." Similarly, there must be another protein molecule to stop the gene from continueing to produce more of the same. Thus, we attributed the on/off switch to these two gentlemen working with parasites that make my intestines their home for life. We know that environmental factors can influence a protein to change its shape to provide a response to an external stimulus like very stressful situations. We also know that stress not only influences our responses through shape changes in protein, but most importantly, chronic stress can kill genes that produce the protein we need in our body. Our 45 years of experience working at a mental health clinic applying scientifically proven methodologies qualify me to write on harm done to a person exposed to chronic stress. There are, at least three areas of the brain that are extremely susceptible to stress: the hippocampus, amygdala and the pre-frontal cortex. These three organs or cluster of neurons are considered by most neuroscientists as part of the limbic system. As you already know, the limbic system is known as the home for emotions. It also happens that emotions are a very important attribute of human beings. Through emotions we socialize, makes friends, enjoy ourselves at parties, go on vacation to enjoy a couple of weeks away from stress at work, and through emotions, we fall in love. Another attribute that separated us from the rest of hominids is language. Through words we communicate our feelings, our emotions and in time, we were able to develop signs to write down our emotions. We keep our autobiography through episodic memories recorded in our brain. Besides, as a family, tribe or nation, we pass on down our emotions in written form from our heroes. Emotions, if expressed in exaggeration, can be overwhelming and drive a person "nuts." When we fall in love it seems that the pre-frontal cortex, our most rational cluster of neurons, is over ruled by the limbic system. Likewise, when anger takes over rational thinking, an individual may commit injury to himself or others.

You do not need us to tell you how a bugging stressor at home, job or neighbor can get your adrenaline to hijack your body and drive you nuts. The same argument can be applied to people predisposed to brain diseases like schizophrenia, depression, heart problems, hypertension, etc. Anxiety does not need a specific trigger to get you upset. We live in a world that is anxiety provoking 24/7. We turn on the radio or television, and stress triggers get to you right through your ears, eyes and skin.

We began this journey on DNA and RNA not only because we are fans of Watson, Crick, Collins and Venter, but because they have helped me understand myself through their work and throughout their professional careers. It has been our main concern to understand the person we see during our clinical practice as a whole entity. The person we see in our office is made of the same elements in my body. He or she is subject to the same daily threat and risk I do. We may ride the same subway, take the same bus to and from work, breath in the same polluted air, and often, eat the same fast food from the same restaurant. We live in cities with police cars sirens, ambulance and fire-fighters trucks alerting us for possible life hazards. Now, we come to a personal experience. We come from a family with a thalassemia minor history. Thalassemia is an inherited disease. There is no cure for this genetic disorder. The disease is classified as thalassemia major and thalassemia minor. I am a carrier of thalassemia minor traits. It is a genetic disease I inherited from my parent's genes. What concerns me most is that about 50 percent of my family, 5 out of eleven children were positive for thalassemia minor. Jack, my brother just immediately before me died before he reached three years old. According to my mother, he died from anemia. My oldest brother, besides being positive for thalassemia minor, died of leukemia after he reached sixty years old. Two of my oldest sisters had children and grandchildren who inherited thalassemia. My oldest sister's grand daughter died of leukemia at age 9. My second oldest sister had her grandchildren follow up at a local health clinic for thalassemia minor. Both, thalassemia and leukemia are blood diseases. Therefore, the suspicion that there may be a connection between these two fatal blood disorders haunted me for some time. In thalassemia, we know that it is a dysfunctional or a mutated gene. However, in leukemia, a blood cancer, like all cancers, the causal factor or factors for this fatal disease is not fully understood. Some studies have implicated genes in its etiology. However, there are many different type of cancer that makes it difficult to point out its etiology to any single gene or group of genes. As we have said earlier, cancerous cells are rebellious cells that proliferate, invade and destroy healthy tissues and organs in our body. The best tools in our modern medical technology inventory have not been able to identify and explain what originated the initial proliferation of malignant tumors cells. There are many endogenous and exogenous trigger factors which make it very difficult to point at a specific or single causal factor. Does breast cancer have the same etiology as leukemia, colon or prostate cancer? People exposed to pesticide, toxic chemical or chronic stress can develop cancer. Can you name people who are cancerous victims to the same etiology?

The Bubble Boy, David

In our case, it was a blood disorder. On thalassemia we do not know if it is a mutation, deletion of genes or a histone modification. We know that we do not produce enough globins to carry this vital element, oxygen, to all parts of our body. Otherwise, our family has been healthy. My father and my mother lived to be almost a hundred years old. There is a movie which pops-up in my brain, the Bubble Boy. It is about a boy named David who was born without an immune system. His body was unable to make antibodies to destroy intruders to his organism. David lived in a sterilized environment (the bubble) until he was 12 years old. A bone marrow transplant from his sister did not help. A virus was present in the cells she donated him and killed David. He did not have T & B cells to alert and organize a host of auxiliary killer cells that would protect him from viruses.

The literature that we researched claims no linkage between thalassemia and leukemia. For thalassemia disease, gene therapy may come with a possible solution. Trials with humans have not produced the desired response, but we were not born flying. Whether we include induced pluripotent stem cells (ipsc), embryonic cells, gene therapy, micro RNA or cloning, definitely we are going in the right direction. We will convert blood stem cells into T cells progenitors and join B cells in the battle against pathogens. Our initial concern over the etiology of leukemia was partially answered during an international symposium in Germany in 2004. It reads as follow: "Several studies reported the derivation of leukemia stem cells from hematopoietic stem cells as a result of epigenetic events...[24] Whether in the U.S.A., Europe or eastern countries like Japan, South Korea or China, embryonic stem cells research as well as induced pluripotent stem cells will be in the forefront of medical research and pharmacotherapy. Under IPSC we can create any cell in our body. It is even possible to build tissue and organs for a body in need of repair. C. Venter's synthetic bacterium may become a handy repair tool. The most beautiful part of IPS technology is that the problem of rejection by our immune system will be overcome because we will be using stem cells from the same person in need of repair. Our organism is a factory of functional cells. Japanese and American scientists are at present time the leading researchers in this field. Making use of these technologies, we will not be doing anything out this world. We have a natural tool factory inside us. Watson and Crick placed us on the road to decipher the human genome code. Now we can enhance our brain to absorb and intelligently use all this technology to improve and accelerate brain cells which have moved us ahead from other hominids.

LEARNING FROM SEQUENCED GENOMES.

What can we learn from Watson's genome sequence and the millions of small nucleotide polyphormisms that are attached to the three billion pairs of the genome? James D. Watson's genome sequence showed about 3,300,000 SNPs or simple substitutes of one base for another at a particular site in the genome. Watson's genome variation, as well as that of Craig Venter, seems to be typical of a genome variation of individuals of predominantly European ancestry. Of the 3.3 million SNPs, 10,654 caused amino acids substitution within the coding sequence. In addition, the

---

[24]- N.Y. Academy of Science, op cit. p.1X

leading author of this article, David A. Wheeler, wrote that his team identified... copy number variations resulting in the large-scale gain and loss of chromosomal segments ranging from 26,000 to 1.5 million base pairs. For an untrained individual alien to genome sequence, these are a lot of base, substitutions, gains and losses of base pairs. Uneducated and unaccustomed to see so many changes within a single genome, it might be a little scary. However, it happened to a Nobel Prize winner and brilliant scientist, J. D. Watson. The point we try to convey is that so many variants did not affect him mentally or emotionally; he is mentally, a well-rounded individual. We have the genome sequence of two more brilliant scientists, C. Venter and Francis Collins. This author has not read enough about the genome of these scientists, but would not be surprised to see similar patterns on it. M.V. Olson commenting on Watson genome wrote: "The genome sequence seems to show that he is a carrier for a handful of mutations...but these mutations have no known effects on him." Watson, along with F. Crick, gave us the double helix structure over half a century ago. Now, these brilliant and remarkable men leave us their genome sequence for us to learn about ourselves through millions of gene alterations found in their genome. We close Olson's teaching lesson with another quotation from him: "The challenge in human genetics now is to learn how to correlate genotype with phenotype (the external appearance) with special attention to disease predisposition and response to therapy."[25]

DNA, SNPs and Disease

A brief comment on the above is not out of order in here. Companies that advertise to get your genetic code decoded and reveal to you where you got your beautiful legs from or how to change the color of your eyes is more of a publicity stunt aimed at profit making than scientific accomplishment. However, there are sophisticated and reliable companies that perform genome scanning that reveal to you potential assets and liabilities in your body genome and you may plan accordingly. Again, a word of caution; interpreting for you the potential assets and liabilities that

---

[25]- Nature, Vol. 452, 04/17/2008, p. 819-20 and 872-876.

may come in millions is another big question. Among those genomic markers are SNPs (short nucleotide polyphormisms). They are DNA variants that in some cases are associated with your susceptibility to a disease. But, of course, because these SNPs show up in your DNA, perhaps in the wrong place, it does not mean that you will get the disease. We like to remind you once more that both, J. D. Watson and C. Venter sequenced genome show liabilities; however, generally, both are healthy and brilliant scientists. Both have been a blessing to humanity. Take for instance, variants of a gene dubbed APOE associated with Alzheimer risk. Are SNPs like structures, also variants that contribute and determine the degeneration on neurons that ultimately end in Alzheimer? How are introns, the non-coding genes, and their variants related to the disease? How does exercise, meditation and a mentally busy and well-focused brain like those of Venter and Watson influence the growth of these variants as a liability or an asset? How these companies interpret the propensity to a specific disease with so many epigenetic and environmental risk causal factors jumping on us 24/7? Besides, a good number of these findings are based on statistical average and there are a significant number of samples that do not fit or fall within the average category. We do not know, but if we were to go by the average sample, let us say in scientific talent, John Nash, C. Venter, J. D. Watson, A. Einstein and even Da Vinci four centuries ago had to be counted out. I am not throwing out the average scale. What I am saying is that before you lose sleep and money over your DNA variants, please pay a visit to your physician or a qualified health professional.

Genome surprises

Genome sequencing has brought us many surprises. We have sequenced a few animals including humans from European ancestry, Africa and China who have provided us with insight into our own evolution and development. Among our surprises was the incredibly large amount of non-coding genes taking space inside our skull. Besides SNPs, epigenetics, mutations, misspelling and others, we have learned that a mutation- a change of a single letter of DNA code- is not that bad at all. During gene duplication some will be ill prepared to function normally and will be degenerated. The cell has its own self-correcting mechanism. However, an occasionally produced mutation may slip by and allow a duplicated gene to prove itself by doing new and useful jobs for the good of the cell, and ultimately, your organism. A gene may strive to serve the cell and may give rise to competition of functions. Smell and sight in dogs and birds might well illustrate our point. We cannot forget that shape determines function. Many things take place during cell specification that we cannot elaborate in this essay. Here, we are not talking about the battle of the fittest, but which shape fits in and serve best the cell and organism. "We humans have about 400 genes for smell receptors alone, all of which derive from just two in a fish that lived around 450 million years ago.[26] This is not a short trip by any means, but there is not a way we can compare our sight and smell with that of an eagle, a hawk or a dog.

During the last decade we have been looking deep and close into the intricacy of the genetic code of life. Letters change within a gene, misplaced codons and introns into promoter's domains, as well as DNA and histones modifications can interfere during protein synthesis, embryonic and fetus development. Consequently, all these steps may provoke malformations and diseases on us. Even a complete chromosome can be duplicated as it happens in Down's syndrome. In 2008, biologist Wen Wang in China, working with several species of fruit fly, an excellent research tool, was able to identify new genes that have evolved over millions of years. To his surprise, he found that 10 percent of the new genes had arisen through a process called retroposition. Retroposition occurs when a MESSENGER RNA copies of genes-this is the RNA template sent to the protein assembly factory, the ribosome. These retroposition copies of genes are turned back into DNA and are inserted somewhere else in the genome. The interesting thing for us in this discovery is that the gene copies created by retroposition are not the same as the original ones. We know that genes consist of more than just the sequence coding for a protein. Among laborers involved in gene and protein synthesis are promoters that will send signals for additional participating

---

[26] -New Scientist, 11/22-28/ 2008, p. 44.

molecules.[27] Promoters, as gene regulators, are something we have to reconcile with. They are involved in determining the location, time and volume of the protein to be made. Promoters are facilitators that we need to fully gain to our side and convert it to first class research and therapeutic tools. My friend, the inquisitor, just whispered into my left ear: What you need is a tool to eliminate most junk DNA. In addition, you will place your limbic system under the strict control of your pre-frontal cortex. You may also want to put together a synthetic pathway between the hippocampus and long- term memory storage loci so you can turn it on and off at will. A synthetic gene in a chromosome, as far as we know, is science fiction at present time. However, gene duplication seems to be as common as mutation which is changing a single letter of the DNA code. During gene duplication and reshuffling of its component parts, the resulting "new gene" may well be a different gene from my own ancestor pool of genes. Genes are the instructions to make proteins. And… proteins are the basic pieces of matter nature made it necessary to build your body. Furthermore, genes hold the secret of your entire life ever since you existed as a simple organism. By studying your genes, you will learn about all the secrets your own ancestor did not share with you.

The last ten years in genome sequencing

A lay person overview of genome sequencing during the last decade ending in November 2011 would include speculation about protein coding genes from over 150,000 to about 22,000.

Junk DNA, epigenetics, and probably, its similarity with other mammals like the mouse and a chimpanzee would not be considered. However, Professor E.T. Dermitzakis, from the Dept. of Genetic Medicine at Geneva, Switzerland, points out that in the genome, the exact location of regulatory regions was unknown. Only a small fraction of variations that exist in the human population had been characterized. The basic components in each genome are largely the same, but the way they are used differs

---

[27]- New Scientist, 11/22-28/ 2008, p. 46.

from tissue to tissue and from person to person. Understanding the rules of gene regulation, the grammar of the genome, is key to the understanding of the human body.[28] Professor Pardis Sabeti from Harvard University said: "During the pregenomic era, revolutionary genetic was a painstaking process. Scientists hypothesized instances of selection and sought confirmation case by case." As of the year 2,000, only a handful of cases had been identified. Technological and analytical advances in the past decade have enabled us to progress from hypothesis testing to hypothesis-generating science...We can scan the entire genome to identify variants under natural selection. [29]

Chief Scientific Officer, Eric Steps from Pacific Biosciences of California said, "The first human genome sequence, published in 2001, provided a canonical reference from which to understand genome structure, as well as registry of functional units...At present, thousands of genes that influence susceptibility to hundreds of diseases- associated phenotypes has been identified. [30] Science fiction is fiction until we learn to do it. I hope I can quiet down my inquisitor for a while.

WATSON, CRICK AND VENTER GENOMES

In general parlance, J. D. Watson's genome sequence showed us that his biochemistry served him well. The proteins, amino acids, codons, genes, DNA and RNA that we have been talking about, Watson left it for us in his genome. They come from a person dedicated to work for all of us. It served him well producing a brilliant scientist who locked his chemistry with another jovial and equally smart person, F. Crick. Crick's chemistry connected very well with that of Watson. Neither of the original players of the double helix structure stuck together as Watson and Crick did. We wonder to what extent Watson's SNPs and mutations, all parts of the dynamic of our chemistry, played a role in his behavior and stamina to stay on course. To succeed, he ignored road blocks encountered during his challenging life. We remember the biography of another outstanding scientist of the

---

[28] - Science, Vol. 331, 02/11/2011, p. 689.
[29] - Science, Vol. 331, 02/11/ 2011, p.690.
[30]- Science, Vol. 331, 02/11/ 2011, p.691.

XIX century, Madame Marie Curie. She defeated prejudice and male chauvinism and remained loyal to her ideal and love for science. Her "chemistry" will endure for centuries to come. She died doing research in her effort for us to understand ourselves and live a happy life. Her laboratory was her universe and everything in the universe was subject to scrutiny and analysis. Cause and effect was the golden rule to do research. Today, we do research with tools that can be observed and replicated in order to make use of our laboratory findings for the benefit of all human beings. One of our most widely used research model in medical research is a mouse. Its gene is very similar to human genes. Most recently, in 2011, an orangutan's genome was sequenced and found that its genome is 97 percent that of humans. We already know that our next closest relative among hominids, the chimpanzee, is 98.4 percent like us, humans. You may wonder how the sequencing of these animals help us medically and otherwise. Just think about chromatin methylation and modification as well as other relevant epigenetic causal factors in human illnesses. They may be clarified through the sequence of animals with a genome close to ours. For instance, when we silence a critical gene in a chimpanzee we learn how it affects human health and behavior. Consequentially, we can develop medical therapeutic tools to either correct the gene and its proteins or arrest the disease.

Gene targeting is not Sy Fy

Gene targeting does not belong in science fiction television shows intended to get your adrenaline running high in your body. Our preferred animal of research model, the mouse, has the same organs as humans do, and their genes are about 95 percent identical as any male or female specimen of our human race. Using gene targeting technology, scientists the world over, can knock out, meaning, turn off, suppress or totally inactivate a gene in a mouse mimicking a human disease. This medical technology can be applied on real human diseases. Scientists can force a mutated gene to reappear at specific organs and tissues in our body, including our brain, and observe its behavior as it relates to a particular illness. The knockout technique has provided genome researchers, particularly, medical researchers, the opportunity to identify and observe each gene in our body from birth to death using the mouse model. Our most recent estimate reported that we have about 24,000 genes in our genome. You may argue that with our modern technology, learning everything about our genes should not be an insurmountable challenge for us. You may add that identifying a disease risk marker in our genome seems to be easy job. This is exactly what we thought when we first started searching for dysfunctional genes. The problem we had not anticipated is the role of junk DNA, DNA methylation, histone modifications, and in general, all epigenetic and environmental agents contributing to the etiology of any disease in our body. There are millions of tiny troubling units in our human genome whose function are unknown. These units and functions are like uninvited guests with weird behaviors. Making this issue even worse, they have made our bodily organs and tissues their permanent home. But thanks to our brain, we are determined to get to the mountain top and get control of our destiny and health.

The science king of the XXI century

Nanotechnology, our most recent brain child, has engineered tools to look into, and possibly repair, not only dysfunctional genes, but also to go into molecular and atomic levels of research. It is the science king of the XXI century. Unless we use it against each other, we will change the way we see ourselves and the world around us. In a quarter of a century of research, we went from a computer that needed a truck to move it around, to something as small as your wrist watch. We have discovered materials stronger than steel, but lighter than a feather. It is electronic propelled by tiny vehicles capable of achieving light speed. Sun energy will be used to energize traveling miniature vehicles carrying human cells with memory carrying proteins. Cells will replicate at sites chosen by humans on planet Earth. Hibernation will be used as an adjunct technology during the voyage. This technology is not in Sy Fy books for recreational purpose, it is in our laboratories. The suspected anticipated problems do not lie in our brains, it lies within our economic, social and political systems around the world. How to feed and control around 10-11 billion people on Earth within the next forty to fifty years is our greatest challenge. A recent article in a prestigious scientific journal wrote that feeding 15 billion people by the middle of the century would be "easily possible." Easily possible if more land is used for food cultivation. However, that land is located in Africa and in Latin America. A problem already existing in those places is not mentioned. Land in Latin America and in Africa that in the past was used to feed people is now being used to produce some sort of fuel for the developed world's upper class to ride their luxurious car, yachts and airplanes.

Communication among people would be much easier and accessible to the majority of the population in any given country. Migration around the world will be an enormous problem most difficult to control. In addition, the world areas for future agricultural use, is exactly the place for very fast population growth. India and China and countries between those two, have about a third of the world population. They will be moving out of their geographical boundaries rather soon. There will not be a power strong enough to challenge it. They are already doing it in many parts of the world. Western Europe seems to be inundated with migrants from Eastern and African countries. They arrived by homemade boats, by walking, by trucks, and as "tourists." There are justifiable arguments that we will come to some sort of agreement for self-control of unrestrained birth. It is a little over a half century ever since we established the United Nations to avoid wars. You pass judgment on its mission. The Korean War, Yugoslavia, Iraq, Afghanistan, Israel-Palestinians and Lebanon have been fighting each other ever since the U.N. was established. The big boys at the UN are pursuing their hidden agenda. Now, imagine how can we get together to solve problems of food on the table for billions for people? Agreeing on a single subject has not been a strong point among human behavior. Man spends more money on defense weapons (which actually is nothing more than a weapon to kill his neighbor) than food, health and housing.

The U.S. Department of Energy has launched a late technology initiative to convert sunlight into hydrogen and other fuels. Its director, Nate Lewis said, "We have to scale up from nanoscale to macroscale." Hopefully, it will capture photons and produce electricity, which can be used to split water molecules and produce hydrogen. With artificial photosynthesis, photons from the sun would drive a wireless chemical conversion process to generate fuel.[31]

Converting land dedicated from millennium to produce rice, corn, wheat, bananas, etc., into fuel for the super rich is not the solution to the above mentioned problem. The answer already exists, and we thank our brain for it. Genetic engineering seeds, animals, fish and plants have been among us for millennium. But, at present time we can do it faster and more abundant for all of us. We engineer seeds to fight insects, and survive during drought. Instead of one crop a year, we make it for two crops a year. We can also improve the size of fruits, vegetables and roots. Most of it has come out after we introduced the double helix and began sequencing genomes from animals and plants. Genetically engineered maize and soya bean has been one of our greatest accomplishments making use of DNA technology. Conversion of fertile land to provide fuel to move vehicles for the rich is not going to solve our world problems. The sun is our greatest source of energy we need to explore, besides, how about hydrogen from our oceans? We will have to invest on research with little initial profit, but the final solution will save us all.

RNA- AMINO ACIDS-PROTEINS RESEARCH

DNA, RNA, amino acids chains and proteins research has kept our mind focused for many years. The double helix comes even in food boxes. How proteins are made was a curiosity we could not put away or silence even during sleeping. Researchers have helped us during this dream trip by discovering the basic RNA molecules involved in protein synthesis. We learned about r RNA, t RNA, m RNA all involved in the making of a protein. Interference RNA is another molecule that becomes a nuisance when it interferes during the synthesis of a protein molecule. We all know that proteins are peptide chains. The above mentioned

---

[31]- Nature, Vol. 466, # 7306, 07/29/10, p. 541.

ribonucleic acids and molecules play a significant role carrying and transporting its components polymers from the cell nucleus to the final assembly station, the ribosome. Recently, Phillip Kapranov et al. in the journal, Nature, 7/29/10, suggested a novel RNA copying mechanism. He claims that RNAs are not merely degradation products of longer transcript, but could indeed have a function. Profiling of RNA, including the sRNAs, can reveal not only novel transcript, but make clear prediction about the existence and properties of novel biochemical pathways operating in the cell. This article appealed to us because, among other things, we have maintained that RNA is our body's mechanic that puts our parts together after receiving the message or order from the cell nucleus. RNA folds itself in many shapes carrying multiple functions that make us show more respect and admiration for this acid than ever before. Some readers may think that we are overly concern about i RNA as a "spoiler" in protein synthesis. Their concern might well be justified. However, we invite you look at pharmaceutical and investor corporations that wants to convert i RNA into a therapeutic tool to take care of some of our illnesses. RNA i was discovered in 1998, and in 2006, the group of scientists who discovered it were awarded the Nobel Prize. Soon thereafter, Merck, the giant pharmaceutical company paid more than one billion dollars in its competition to make drugs out of i RNA. Another giant in the drug business, Roche in Switzerland gave up in this competition after spending more than ½ billion U.S.A. dollars on the technology. Furthermore, please listen to Heidi Ledford's writing for Nature: "The development of RNA i based drugs has stalled as companies confront the challenge of delivering RNA molecules, which are notoriously fragile to target cells in the human body, and then coaxing those cells to take up the RNA." There is another quotation to add. "Getting these molecules exactly where we want to go is a little more difficult than originally thought." added Michael French, chief executive of a Biotech company in Bothell, Washington.

Equally curious and interesting for us is our body building block, the protein. There are some proteins and their receptors that get a lion share of our attention. Naively, our attention has been geared at our brain cells and membrane receptors. Heidi Ledfor writes: "About one quarter of approved drugs target members of a single protein family, the G protein coupled receptors." She further added: "Members of this clan- the largest protein family in the human genome, control everything from hormone signaling to the perception of light and scent. A no small amount of money from the U.S.A., $290 million will be re-directed from learning amino acids folds that comprise a protein…to solving some of the world most troublesome and medically relevant proteins including the G protein coupled receptors." Besides this protein, mitochondria as well as proteins that regulate gene expression will be principal targets. We feel relieved by this announcement. The focus will be on proteins, and specifically, proteins that are medically relevant. We want to know not only how proteins are formed, but how they relate to and become the causal factor of many of our health problems. You already know that protein packaging provides for it shapes, functions, as well as the length of the chain of peptides. In addition to multiple containing subunits, it provides for its different roles in cells, organs and tissues of our body.

A late word on protein synthesis before we pass on to the next paragraph is necessary. The mantra or dogma among biologists is that a protein needs a structure to be able to function properly. Every book I read emphasizes that function demands structure. There have been serious arguments that even the most unstructured ambient is ruled by some type of structure and order. Our personal opinion is in consonant with structure, function and order. Peter Wright, a protein biophysicist at the Scripps Research Institute in La Jolla, California, told a group of scientists attending a meeting of the American Association for the Advancement of Science in Washington, D.C., "The recognition of disorder has grown dramatically." Data are fast accumulating from all fronts-biophysics, bioinformatics and cell biology in support of widespread of disorder. P. Wright and team member J. Dyson back in 1999 wrote a review pointing to the growing collection of proteins that seemed to function despite their disordered state. However, there are some researchers with a broad imagination like Joel Janin from Gif-sur-Yvette in France who says that "the whole concept of disorder seems incompatible with the lock and key model. You might as well try to open the door with cooked spaghetti." Little by little, a fundamentally new structure of

71

relationships between protein sequences, structure and function is emerging. It consist of a continuum running from the most rigid lock and key enzymes and molecular machines at one extreme through to durably unstructured spaghetti, spanning all degrees of structural ambiguity in between."[32] How this cooked spaghetti structure will work or is already functioning, is an enigma for us, but we will keep an open mind and see how it turns out to be. When Rosalind Franklin, J. D. Watson and F. Crick were attempting to put the double helix together, doubts seemed to outweigh good intentions.

G proteins receptors

What are G proteins-coupled receptors? It is a large group of proteins that are attached in the plasma membrane of most cells. They seem to look for, and attach to agonist molecules responsible for conveying sensory information. This receptor is also involved in the transmission of changes in physiological states of the cell. Once the receptor is activated, G protein initiates signal cascades that control multiple biological processes. How is it best to approach this challenge is going to require elaborate team work. The mouse and the petri- dish are excellent research tools, but translating its finding to a protein receptor physiology is another issue. The cell internal aqueous environment is further complicated by internal chemical reactions. Likewise, the cell's outside is equally influenced by an aqueous environment. Protein shape and function are conditioned, among other things, by its aqueous environment. How water molecules come in contact with proteins in different cells, affects proteins functions and related medical problems? The dynamic exchange of water and protein atoms and molecules exchange in humans at different stages of development is another challenging issue. This huge family of protein receptors is responsible for many functions we are not equipped to fully elaborate on. However, missing a chance to research the most recent and relevant literature available is something we could not avoid. This is not a local issue, but a worldwide challenge.

---

[32]- Nature, Vol. 471,) 3/10/2011, p. 151-53.

A G protein coupled receptor is essential for Schwan cells initiate myelination. Signals from the axon activate expression of specific transcription factors such as oct6 and kro20. I write these names and numbers just for reference purpose, not to bug your brain. These two molecules begin myelination in Schwan cells. These are glia cells that produce myelin in the peripheral nervous system. Hold on, I will explain it for you. Myelin is a lipid or fat that is wrapped around the axon. It insulates the axon and allows the electrical message (yes electrical message) reach its target fast and uninterrupted. Lack of myelin in the neuronal axon can be fatal. It may provoke serious illness such as multiple sclerosis, amyotrophic lateral sclerosis and Krabbe. A type of brain cell called glia oligondendrocytes produces myelin. In your brain you have cells known as neurons. Neurons are composed of a body which looks like a round disk with many processes or fibers extending out. These extensions are called dendrites and receive messages from other brain cells, in particular, neurons. There is another process or fiber coming out the cell body called axon. Messages from the neuron nucleus go through the hillock into the axon. A hillock looks like a ring linking the cell body and the axon. The axon is wrapped around with myelin almost like plastic is wrapped around an electric cord or wire. There is a difference between the axon wrapping and the plastic covering an electrical cord. First, the axon is not totally covered by myelin. There are small segments or spaces on the axon that are naked or free of myelin. These naked spaces allow electricity in the axon to gain speed jumping from one node to another down the end of the axon. This electrical message or impulse will provoke a chemical reaction, meaning, a neurotransmitter will be released at a synapse. A synapse is a space between each neuron in the brain. Among the neurotransmitters released into the synapse are serotonin and dopamine that we mentioned on above pages. Loss or destruction of myelin most common in the central nervous system is in the brain, spinal cord and optic nerves. Multiple sclerosis is the primary disease of demyelation. Some of the diseases caused by destruction of myelin that we know of are: multiple sclerosis, Krabbe Tay-Sachs, phenyl-ketonuria, Hurlers syndrome and neuro-myelitis optica. Of the above diseases related to myelin loss, I have had professional experience with multiple sclerosis and Krabbe.

M.S. is characterized by recurrent episodes of demyelation. It is a chronic autoimmune inflammatory and neuro-degenerative disorder of the central nervous system. The duration of each attack and how often it takes place do not seem to be well established as yet.

G protein coupled receptors are responsible for the majority of cellular responses to hormones and neurotransmitters as well as the senses of sight, olfaction and taste. The B_2 adrenergic receptor has been a model system for the large and diverse family of G protein-coupled receptors. During the last 10 years, sequencing of the human genome led to the identification of over 800 GPCR genes. The B2AR-Gs complex crystal structure provides the first high resolution view of trans-membrane signaling for a GPCR. Rasmussen et al. have a frame work to design experiments to investigate the mechanism of complex formation, GTP binding and complex dissociation.[33]

Dies Meijer, writing for the journal, Science, says: "Myelin is laid down and maintained by dedicated neuroglia cells-oligondendrocytes- in the central nervous system and Schwan cells in the peripheral nervous system."[34] My experience with Krabbe was a very sad one. A little girl 18 month old otherwise considered healthy by the local pediatrician began crying without any obvious provocation. Her mother took her to the local physician, but to no avail. She was later taken to a specialist who diagnosed her as Krabbe disease. She was later transferred to a well-known pediatric hospital in the East Coast of U.S.A. Her initial symptoms were: feet turned inward, unable to stand up, could not walk, and complaining of pain all day long. Soon afterward, the little girl began to lose her sight and hearing. After a few weeks in the hospital, she was discharged home to her mother's care. She is in a wheel chair with weekly nursing visits and monthly physician checkups. Another reliable resource person on myelin and Schwan cells wrote: " The axon of many vertebrate neurons are enclosed by a chain of supporting cells called Schwan cells in the peripheral

---

[33] Nature, 09/29/2010, p.549-555
[34] -Science, Vol. 325, # 5946, 09/11/2009, p. 1353.

nervous system, that form an insulating layer called the myelin sheath."[35]

A personal encounter with histamine

During my clinical practice I have seen many people with rare symptoms and diseases that have strengthened my dedication to assist individuals in pain. One area that always challenged my ability to help anyone is allergies. We have been fighting allergies for a long time. From dust, pollen, weeds, cats, dogs, food and, even an occasional rare perfume would provoke a wheezing sound, running nose and watery eyes. A visit to a physician usually brings relief after taking the prescribed medication. However, the prescribed medication does not always work. We should not blame the medication or the physician. Although we are made of 20 amino acids that will be formed into proteins, the end product of amino acids and proteins, your body and mine, are not truthfully equal. We end up being two unique individuals responding to internal and external stimuli in different fashion.

However, despite our differences in behavior, there is always an average response. And…a medication or rather the usefulness of any medication is measured on the average individual, not on the difference between each individual person. Consequently, a medication that worked wonders in Peter will not necessarily make the same miracle for Joe or John. Remember, we carry our genes from our ancestors, but we are also greatly influenced by epigenetics and our immediate environment. My mother and two of my siblings were always complaining about allergies. Allergies have been my inseparable companions for many years besides having a relatively poor digesting system. Greasy foods like pork chops, milk, cheese and even butter pecan ice cream that makes my mouth a lake of saliva, do not fall right inside my stomach. Somehow, those protein-enzymes responsible for breaking down fat are not helping me. I avoid those foods although my eyes and an internal voice called me a coward for ignoring my desire to enjoy myself. From the health food store I

---

[35] -Biology, 3rd Ed., Neil A. Campbell, the Benjamin/ Cummings Co. N.Y., p. 984.

buy appropriate enzymes. Besides allergies, I have learned about another health nuisance, histamine. We have said that our organism is a very complex chemical reaction factory producing millions of reactions that we are committed to fully understand. You are correct when you say that many of those chemical reactions are the effectors of some of our diseases. When you experience an excess of gastric fluid in your stomach and feel that bitter taste, you know what it is that I am talking about. Health foods stores and over the counter medication at pharmacies have full shelves waiting for you to choose from many similar remedies. To calm down my stomach acid and running nose, I carry with me the antihistamine called, Diphenhydramine.

Histamine, a chemical messenger

Histamine is a chemical messenger that is responsible for handling a wide range of cellular responses. We have been talking about DNA and RNA molecules going through multiple reactions until proteins come out to form organs and tissues. Take a minute and visualize all chemical reactions going on in this precise moment inside your body, and specifically, inside cells. Among the reactions we are referring to are allergic and inflammatory reactions. Most likely, you must have had an allergic reaction of some sort at one time. What is not visible and difficult to explain is the inflammatory reaction taking place inside you. Some asthma attacks could be an allergic reaction located in the lungs. Besides or along the inflammation on your lungs, you may have an excess of mucus that makes it very difficult for you to breath. You feel like choking on your own saliva and mucus. You cough and cough again and again to clear your throat and breathe freely. In addition, and to my surprise, there is an excess of gastric fluid in your stomach threatening your health. You try to cough out all the mucus coming out your lungs, but from inside your stomach comes out a full cup of vomit. It is as bitter as nothing you have tasted before. All this pain and discomfort was provoked basically by histamine or an excess of it. This chemical messenger mediates a lot of responses in millions of cells in your body. It is suspected that it works in brain cells too. Besides all the problems that an excess of histamine can create for you, it is a powerful vasodilator.

We do not chicken out because some illnesses are hard to crack down. Take for instance, serotonin; it is highly involved in brain functions. It is a neurotransmitter involved in depression. However, the largest amount of serotonin is found in cells in the intestinal mucosa. "Serotonin has multiple physiologic roles, including pain perception, normal and abnormal behavior, including affective disorders, and regulation of sleep, temperature, and blood pressure."[36] Besides our well known serotonin, we have another public friend that most people talk about even over the dinner table, dopamine. This neurotransmitter is a chemical that make you feel extremely happy when produced in excess quantities. It is also involved in two chronic and degenerative brain diseases: Parkinson and schizophrenia. In Parkinson disease, cells producing dopamine in the brain substantia nigra die out. Recent medical technology is helping people with Parkinson. In schizophrenia, there are some genes involved in it, but this disease is more of a syndrome than a single encompassing behavior. There are effective medications available at present time to control most of the symptoms of schizophrenia. Schizophrenic symptoms are not difficult to identify by health professionals. Youngsters from ages 16 to 21 are very vulnerable to strong psychological and physiological stressors. These individuals with a schizophrenic pre-disposition need extra help from parents and teachers as soon as possible.

E. KANDEL RESEARCH ON...

Dr. Kandel and Eleanor Simpson, both neuroscientists at Columbia University in New York City, reported in 2006 a discovery made on D2 receptors associated with schizophrenia. Dopamine receptors D2 had been implicated in the etiology of the disease. These two scientists used engineered mice to mimic the deficits of short–term memory and attention in schizophrenia. The schizophrenic mind loves to travel. It is jumping from one theme or subject to another, unable to focus in the here and now present time. His brain capacity for hallucinations and delusions seem to

---

[36] - Biochemistry, P. C. Champe & R. A. Harvey, 2nd ed., J.B. Lippincott, Phila. 1994, p. 265.

be a bottomless pit even during his catatonic state. The dopamine receptors D2 suspected of contributing to schizophrenic symptoms are located in two region or groups of cells deep in the brain. These two clusters of brain cells-neurons- are commonly known as striatum, consisting of putamen and caudate. They are part of the basal ganglia which are associated with another chronic and degenerative disease called, Parkinson. Dr. Kandel and his team engineered mice with at least 15 percent more D2 receptors in the striatum than normal. The researchers found some deficits for working memory. A working memory is of a short duration; it is for you to handle the task at hand. It is a here and now memory. Do not bring in synaptic sensitization and permanent anatomical neuronal changes by protein synthesis. They do not fall in this group of memory. The mice in question took much longer than control mice to master the task or game during the experiment. Control mice were not engineered; their D2 receptors were not touched. The researchers concluded that extra D2 receptors cause permanent brain damage. Dr. Kandel added that damage probably occurs while the brain is still developing. This experiment dealt with schizophrenia symptoms of learning and memory only. It did not deal with hallucinations, delusions, mood and occasional erratic and violent behavior of schizophrenic patients. We will follow up on this experiment as we explore the brain. An interesting observation we like to point out is the connection with the developing brain. There have been other researchers who have followed up attention deficit hyperactive disorder that have turned into full blown schizophrenia. In addition, from parents, teachers and mental health professionals dealing with children behavior, we can observe early symptoms. These observations and identification of early symptoms can contribute to early diagnoses and treatment for schizophrenic patients. During early adolescent hood while the pre-frontal cortex is going through immense changes, dendrite growth and maturation may relate to Dr. Kandel's observation that the changes he noted occur while the brain is still developing. Dr. Solomon Snyder at John Hopkins Medical Institution in Baltimore concluded that the mice findings reported in the journal, Neuron could help researchers develop new drugs to prevent such early brain damage from occurring in people who may be susceptible to schizophrenia. Whether I like it or not, luminaries like John Nash,

Salvador Dali and von Gogh popped- up in my brain. They contributed so much to humanity despite their inner pain, that we would be a lot poorer without them. Schizophrenia has been an enigmatic disease for millennium. It has been blamed as a punishment from heavens, and/or a curse from relatives and friends. Schizophrenics have been ostracized, imprisoned, burned at the stake or even guillotined. Most of our medications for schizophrenics is aimed at controlling its chemistry, like dopamine and serotonin. The big problem we face is that it is a syndrome rather than a single isolated symptom you can easily identify. The schizophrenic patient goes from a complete flat affect to an emotionally hyperactive individual. During the flat affect he or she may be verbalizing to you the most horrible accident his family has experienced without implicating an ounce of emotion involved in it. During the therapeutic session, you may wonder if the emotional part or parts of his brain were surgically removed from the person you have in front of you. However, the same person seemingly without emotions, may become agitated or emotionally aroused by a sensorial stimuli or a past episodic memory, and present to you a seemingly normal affect. His mood may change from time to time unpredictably. History records of mankind show that we have had the disease with us ever since we began to have symbols describing our behavior.

Alois Alzheimer and spaghetti strings

Before we proceed with our beloved proteins we would like to write down a few words about one of the most dreadful dementia of all, Alzheimer disease. We mentioned it briefly in another paragraph. Alzheimer disease was first fully described by Alois Alzheimer in the XIX century and continues to be as frightening and elusive as ever. We know that spaghetti string like proteins or peptide chains known as plaques and the protein tau are, as far as we know at present time, the two main villains or culprit of this degenerative and chronic brain disease. This is the disease that robs you of your personality because it destroys your memories. These plaques block communication between neurons making synapses disappear. A synapse is the space between neurons. A chemical neurotransmitter is released at a synapse. This chemical message will bind to a receptor at a post-synaptic neuron. Another destructive scenario takes place when the neuron's internal organelles or fibers break down and form an entanglement that collapses the neuron. The complex and harmonious functions of each neuron cease to exist. What is left of a healthy human being is a body with a head; but the brain inside the head is no longer a chemical and electro genesis machine. The neurons that normally made it possible for you to do complicated and elaborated tango steps, piano playing, do surgery or appreciate Picasso and van Gogh paintings are not available to you anymore. Even when whispering at your ears sweet voices from loves ones, there are no receptors in your brain cells to respond to and reciprocate the sweet feeling. This disease is generally associated with old people, but we have had patients in their forties. It is not an old people only disease after all.

Six years ago, (2006) William Dagust, a physician and neuroscientist said at the time, "Many people believe that we are going to have treatment for Alzheimer in five years." Many miles away in Sweden, a group of researchers tested the spinal fluid for concentration of tau protein and beta amyloid, both strongly implicated in this disease. The group tested consisted of 137 people ages 50 to 86 years old. Patients whose spinal fluid at the beginning of the test had abnormally high tau protein and low beta-amyloid concentrations were nearly 18 times more likely to develop Alzheimer as compared with normal concentrations. Those researchers had hope that there would be effective drug treatment by 2010. There is a lot of confidence in our scientists and technology most recently developed. An effective treatment, in our opinion, is not far away. We are joining hands and resources to defeat this disease.

Following on proteins

No other molecule in the living world, except DNA, is chemically, structurally and functionally as complex as proteins. Although the cell nucleus sends out the message for protein synthesis, the message itself, (meaning the selection of specific nucleotides that will recognize a complementary molecule with an amino acid to be pasted together at a ribosome), is taking place while other chemical reactions are in the cell's cytoplasm, organelles and tissues, is an awesome engineering project accomplishment. Crucial among choices to be made (if it is a choice) during this chemical wonder, are selections of codons, appropriate amino acids, often within a group of six triplets, amino acid group, carboxyl group, side chain, shape and function as well as size of each protein molecule. How many peptides for a specific protein function? How do proteins respond to multiple internal and external stimuli and still maintain its integrity and function? Your organism makes many protein molecules with highly specialized functions that are vital for the survival of every living cell in our planet. Imagine a protein that exists in a cell or organism in a dry dessert and compare it with a protein in a cell in the coldest region in Siberia or North Pole. Consider a protein that is involved in generating electricity in a brain cell and one in your right foot toe. There are proteins that are turned into highly skilled soldiers of our body like T and B cells of our immune system. I have proteins that are at least over 81 years old and still kicking. Most of my brain neurons are functioning pretty well. Imagine how many hundreds, if not thousands, different types of protein are doing a specialized job in your body at present time while you read this book. Each one of these proteins has its own particular amino acids sequence. Another educational memory- protein- that we must keep at the tip of our fingers is the property of each amino acid. There are polar and non-polar amino acids that are decisive in protein folding.

From mind speculation to brain research

Just a couple of decades ago, mental health practitioners attributed brain disorders as just mental problems. The issue we want to emphasize here is that 'mental' is considered as having no matter. Mental and psyche was, and still are considered by some quarters as something separate and apart from anything having flesh. By considering mental problems something that we could not assign to a brain tissue or organ, no loci in the brain was considered specifically responsible for any disease in particular. Psychologists ruled out brain tissue and organs as the subject of research and exploration in their efforts to find a reliable and trustworthy cure for problems residing in our brain. They reasoned out that if mental problems did not have biological etiology, it had to be caused by environmental factors. They were not too far off in their efforts to find a culprit for psychiatric diseases and disorders. However, they chose our homes as the first and most relevant factor leading to brain problems. And, at home, all the blame for all types of behavioral problems was attributed to mama, the mother. Second in line in the culprit list was the father followed by family disputes, poverty, etc. I recall some mental health practitioners blaming the mother as "an emotional refrigerator or an emotional armadillo" for her child's autistic, schizophrenic or schizoid behavior, ignoring a biological base for those brain diseases.

Nowadays, scientific approach based on genomic studies, neuroscience, as well as all biological related medical diagnostic tools such as PET, FMRI and similar technological medical tools are changing how we look at and treat psychiatric patients. These machines show us that new brain cells are born every day in specific brain regions. Besides, recent scanners show us abnormal connections between neurons, most likely, secondary to a gene mutation. We have identified genes involved in schizophrenia, autism, Alzheimer and other brain diseases and disorders. We know that brain diseases have a biological basis and are not just "mental" conditions that can be cured by a hand tap on the patient's shoulder. We need to go after genes and its twin brother, proteins in its multiple shapes and functions. In the above pages we described for you how RNA and amino acids are involved in protein synthesis. You also know that proteins are a major component of every cell in your body. Proteins are chains of amino acids. You also remember that many amino acids appear in nature, but only twenty are used by our body. Equally important is remembering that shape determines function, and when proteins take the shape and function of an enzyme, enzymes are responsible for many biochemical reactions. So, proteins come in different shapes and sizes, from just a few amino acids that will form a chain comprised of hundreds of amino acids. We hope you can visualize how complex this protein synthesis is. Let us imagine a protein composed of a thousand amino acids all packed tightly so it does not get entangled with other amino acids forming another protein or other intracellular subunits. It seems that the job that lies ahead of us is not a piece of cake. The beauty of the job that lies ahead of us is that we love challenges. Working on proteins that are particularly relevant to medical research and the solution of psychiatric diseases I have seen is an incentive I cannot fail to celebrate. You can help by attending seminars and workshops provided by researchers throughout our country.

Our basic carbon chips

Recounting our story, you and I are basically made of carbon chips as a basic functional unit of multiple cells. Carbon, oxygen, nitrogen and hydrogen comprise about 95.6 percent of all the elements our body needs to survive. A carbon atom is a very dynamic atom. It makes many decisive bonds with atoms and molecules of other elements that are necessary for all living things in our planet. Carbon atoms are involved in amino acids molecules. All amino acids molecules must have a carboxyl group and amino group, and both groups are joined to a carbon atom. The carbon atom that these two groups of molecules are joined to is known as a-carbon. You know that the basic unit of your entire body, whether it is a hair on your head, a toe nail on your left foot or the upper lip of your mouth, it is made of proteins. And… proteins are long chains of peptides that in turn are made of amino acids. These protein chains coiled up as if it were a ball and therefore are known as globular proteins. So, proteins are polymers of amino acids joined head to tail in a chain that could be several hundred units long. It is folded in a three dimensional structure. The union or bond between two nearby amino acids is known as a peptide bond. The carbon atom is always involved in everything that is cooking in your body. Just think about it, of the twenty types of amino acids found in all proteins, no matter which amino acid is chosen to form a polypeptide chain, it must have a carboxyl group at one end, known as c-terminus, and amino group at the other end, known as n- terminus.

The beauty of all amino acids is that they repeat over and over again in all proteins. No matter what part of your body you begin to take apart for analysis, you will come across with some of the same old good amino acids. You may play around with it adding growth or degrading factors, you will end up with amino acids. The funny part of this game is that you will be learning about yourself. Even before you opened this book, the image of DNA and RNA came to mind. These two acids plus carbon atoms, proteins and amino acids are all about you. Later on, you became acquainted with the famous letters making up the two famous macro-molecules of the two above mentioned acids. The letters we are referring to are: t-a, c-g for DNA and a-u, c-g for RNA. Please, remember these different sets of letters for these molecules.

We have also used the word nucleotide quite loosely, but what is a nucleotide besides the usual connotation of a base? A nucleotide is in itself, a molecule, which in turn, is itself made of atoms. More carefully defined, a nucleotide is a molecule formed of nitrogen containing ring compound that is joined to five-carbon sugar. Basically, there are two types of nucleotides that we like to work on. Nucleotides that contain ribose are known as ribonucleotide (RNA). Nucleotides that contain deoxy-ribose are known as deoxy-ribonucleotides. These are long words that soon became shortened in our daily conversation. In general terms nucleotides are referred as, or named after one of the bases it carries. That is why you read a thymine nucleotide although it is implied that it carries one or more phosphate group. Yes, this is the phosphate- sugar backbone that Watson, Crick, Wilkins and Rosalind Franklin were struggling with while trying to come up with the double helix structure. Were the bases best hanging outside the backbone, or inside making hydrogen bonds? One nucleotide you will not be able to ignore while studying the two great molecules, DNA, RNA nucleotides, is adenosine triphosphate or ATP for short. It is an important energy carrier molecule involved in many chemical reactions inside our body.

From transcription to proteins

Messenger RNA can leave the cell nucleus as a single strand and travel through the cytoplasm and land at a ribosome. This protein assembly factory also receives another RNA single strand known as transference RNA. This molecule reads the message- three letters- that the messenger RNA clearly displays and the ribosome pastes it together. After it is pasted together by the tiny globular molecule ribosome, it is kicked out as a peptide. Multiple peptides form a long chain that we know as proteins. A messenger RNA lands at a ribosome if another RNA molecule does not interfere along the way. Another RNA molecule may block messenger RNA from landing at a ribosome in the cytoplasm. In terms of space, time and size, it is a long distance to go from one place to another. It is a long distance when we consider that it has to go through many smaller organelles each busy doing their assigned job. Besides, there is another RNA molecule known as interference RNA that bonds to target sites on a messenger RNA molecule altering its shape and function. The vast majority of us are blessed from the very beginning of conception if we ponder on the complexity of life as it grows from a zygote to a fully developed and healthy human being. It is not just interference RNA that may interfere with the formation of proteins, but also, micro RNA may do it too. There is a complete family of RNA that may interfere with protein synthesis. The first micro RNA gene was discovered in 1993; five years later, RNA interference was discovered. This is the RNA molecule that made me grow goose bumps when I began to understand its potential power and intervention in protein synthesis and gene activation and suppression. Nicholas Wade wrote for the New York Times on 6/21/05 that I RNA is "a system for silencing genes by tricking the cell into destroying the gene's messenger RNA before it can generate its protein product...Together with transcription factors, micro RNAs may play a role in cell differentiation, the formation of many specialized types of human cell from a single generic type...Micro RNAs create an environment, also tailored for cell type, in which some kinds of messenger RNA can flourish and others are diminished or repressed." Nicholas Wade used the word, tricks in reference to I RNA's job.

We are trying to create for you a picture of the roles ribonucleic acids play in protein synthesis. We had high expectations on our ability to make research tools and drugs from distinct forms of RNA molecules. However, RNA has proven a hard bone to crack as above mentioned companies have found out. We are dealing with the oldest bio-acid we know of. When DNA showed up on Earth, RNA was an old man with gray hair. In the same paper and date as above, Brian Libby wrote: "Dr. Kenton Gregory at St. Vincent Medical Center at Portland is one of a handful of researchers working to create replacement tissue from a naturally occurring protein, elastin." Elastin is known as a matrix protein in that it holds the cells together into tissues and provides a natural support and flexibility. It gives skin and blood vessels their elasticity... Another matrix protein is collagen which is for the tensile strength of tissues.

Proteins, memories and Eric Kandel

For many of us trying to understand ourselves, Eric K. Kandel's brilliant work and discovery of anatomical changes taking place at synaptic sites when permanent long term memory is established in our brain is awesome and inspirational. To our delight, once more, proteins synthesis takes first place with all the complexity of transcription, translation and peptide formation already discussed on above pages. Commenting on a discovery on drugs that inhibit the synthesis of proteins in the brain in addition to disrupting long term memory if it is given during and shortly after learning, short term memory are not disrupted, said Louis Flexner at the University of Pennsylvania. This finding suggests that long-term memory storage requires the synthesis of new proteins. Now, a very interesting and crucial question arose among all of us. Where does the new protein grow in a brain cell during the formation of long term memory? Does it grow on the same site as short term memory? Is long term memory an outgrowth of short-term memory? Is there a specific region or cluster of neurons in the brain responsible for the formation of proteins for long term memory? Doctor E. Kandel has given us answers to our questions. His research at Columbia University provided him with a live tool he used wisely and fruitfully. He wrote: "Short term memory produces a change in the function of a synapse, strengthening or weakening preexisting connections; long- term memory requires anatomical changes. Repeated sensitization training causes neurons to grow new terminals giving rise to long term memory." Neuronal connections already exist, you activate them through a learning experience. Let's say when you learned the first three tables of multiplication. You enjoyed it and you kept repeating it several times. You even went to the next step of learning the six table of multiplication. You felt great; you enjoyed the "feeling good" that we experience when we accomplish something that satisfies us and the loved ones around us. You go out and play ball, play with your dog and even watch television for a short while. You can easily, without any effort, recall all the multiplying tables you have in your brain as long term memory. This type of memory requires protein synthesis, proteins that will be attached to or seen as a protrusion coming out of dendrites. It is as if a new small branch sprouted out a neuron. This is what E. Kandell called anatomical changes in brain neurons. A note of clarification, this is happening in neurons. The brain also has glia cells, but at present time, its role in memory formation is not clear yet. Another teaching lesson from Kandel is the following: "The plasticity of the nervous system (the ability of the nerve cells to change the strength and even the number of synapses) is the mechanism underlying learning and long term memory."[37] Among memories, there is one I have never forgetten when first I entered the physics classroom, $E= MC^2$. Albert Einstein was such an outstanding figure on science that the professor at the University of Puerto Rico had his picture covering the main door as we entered his classroom.

Learning processes have been the subject of study, debate and speculation for millennium. Methods of teaching have been developed by many outstanding educators throughout the ages. Some teaching methods have survived the challenges of time, place and needs of society with slight revisions while most lagged behind and disappeared. How to pass on our body of knowledge -our wisdom- to the next generation is a great challenge to our society today. Equally challenging is the content of the body of knowledge we want our children, and our children's children to learn in order to survive in a highly competitive society. In the year 399 BCE in Athens, Greece, the greatest philosopher the western world has known, Socrates was forced to drink hemlock, a poisonous herb. He was accused of perverting Athen's youth by unorthodox methods. The debate is even more challenging and controversial than ever before. We have issues like creation and

---

[37] - Eric K. Kandel, op cit, p. 218.

evolution, science and humanities including art and history. Some individuals are advocating reducing the last two to a very minimum of time while science and technology should be raised to first priority in a school curriculum. Is education a State responsibility or should it rest primarily within the family? Should education of our children be left to educators and parents? Should politicians be in charge of our schools system as they are in charge of legislating for all of us? The XX century left many scars on us under Nazis and Communist regimes. We are learning rapidly how learning and memory take place in our brains. Sensitization and permanent neuronal structural changes can be easily manipulated in the classroom. These are hot issues that need our attention before it is imposed on us.

More on memories and learning

Once we had grasped the idea on how learning and memories get permanency in our brain cells, we could not stop searching. We turned to Dr. Kandel's engineered and exceptionally successful experiment on a manageable research toy, the aplyssia. We were familiar with short term memory also known to many of us as working memory. We were familiar with H. M.'s surgery on both temporal lobes and removal of most of the hippocampus. After surgery H. M, could not store memory for long term use. All his memories after surgery only lasted a very short time. Somehow, functional memories could not find their way to other brain cells for storage and future use. H. M could not transfer to a cluster of neurons for permanent storage after having eaten a sirloin steak with baked potatoes, French salad and a glass of wine. Soon after he finished his delicious dinner and walked into the crowd on the street, you may want to ask him how he enjoyed his steak dinner. Looking surprised by your question, he would simply reply, "what steak dinner are you talking about?" The steak dinner was inside his stomach, but his brain did not register it. You may wonder if brain cells around the hippocampus in the temporal lobe could be generous enough under the plasticity theory and take over memory storage for H. M. We know that brain cells often take over functions not originally theirs. The physician who took care of H. M.'s health for many years reported that he never recovered his past memories after surgery. Related to the hippocampus, Laura Beil writing for Science News said: "The hippocampus encodes and prepares new memories for storage, then dispatches them to different parts of the brain. In 1989 scientists reported evidence that the human hippocampus is not only a depot for memories, but also, a birthplace for neurons-thousands each month." The nursery for nerve cells is restricted to a raisin-size region of the hippocampus called the dentate gyrus. Supporting the above statements, Fred Gage, a neuroscientist at the Salt Institute for Biological Studies in La Jolla California said that at any given time, about 3 to 5 percent of the cells in the dentate gyrus are in some stage of growth.[38] He further added that most of the dentate gyrus is formed after birth; a lot is formed during the first four years of life. This is very important to know as it relates to growing children during a critical period for learning and forming new memories. Parents and educators, especially early childhood educators, should take advantage of this information to plan teaching lessons because we are talking about one of the main regions of the brain engaged in learning and memory formation. These are memories that will mold our personality for life. For instance, the earlier you expose a child to another language or languages the better. The same is true for mathematics, science, astronomy, etc. Neuronal circuit and pathways as well as synapses are established and strengthened for adult life challenges.

Memories, brain and learning

Whether we like or not, we are, for the moment, stuck with proteins. You know that your wedding date was embedded in your brain as a protein. Therefore, we will continue our inquiry into proteins and memory formation. I carry toxic episodic memories in my brain that pop-up in my mind without an invitation. Likewise, I carry sweet and educational memories that make my life happy. Our interest on protein synthesis, memories and disease is not just an intellectual exercise; it is a very personal one. Besides inheriting thalassemia traits from my parents, I am short sighted, and when I come out of a dark room and face light, I become almost totally

---

[38]- Science News, Vol. 179, # 3, 01/29/2011, p. 23-25

blind for a few seconds. Of course, I wear corrective lenses. I am not an albino person. I have been examined by an ophthalmologist and he told me that I lack some vitamins and tissue in my eyes. Sun rays blind me temporarily. My eyes are slow in adjusting from darkness to light. Some researchers have linked it to the pineal gland deep in the brain. This tiny cluster of cells is responsible for releasing melatonin and is also involved in the wake-sleep cycle. It seems that probing my brain is not a curiosity at all, but a necessity discovering myself through my strengths and liabilities.

We have repeated over and over again that the DNA molecule nucleus is the command center for everything that takes place in the cell. However, this seems to be partially true. For example, strengthening or weakening a synapse is more like a localized issue than a command from the cell nucleus. For instance, if I do not repeat the order and names of planets in our solar system several times and read it aloud, it will vanish from my brain. Someone whose name I do not remember said: "Use it or lose it." Our brain has trillions of possible connection among dendrites. It can store many things, but it is my belief that the brain does not waste energy in trivial things that it does not use regularly. The following might not be a good example, but I will write it anyhow. Coming from my home in the countryside I rented a room where an elevated train passed by every half hour. The first nights were torturing nights, filling my ears with strips of cotton trying to go to sleep. Within six months, I did not see nor hear the train passing by anymore. The cotton in my ears, the train and the noise had gone forever. Yes, it is call habituation. My brain has survived thousands of years, and I hope it can survive many more. My brain uses synapses as needed; it also closes synapses not in use. It is the law of conservation of energy.

CREB protein

We had read about protein kinase A traveling to the cell nucleus for the purpose of activating a gene or genes. In addition, we had learned from Paris that Jacob and Monod, the E. Coli experts, had discovered that messages from a cell's surroundings can turn on genes' regulatory proteins. We were assuming that a particular placed protein needed in a particular place was synthesized by the relevant gene or genes. However, in all honesty, E Coli has never been in my list of research models. Just to think that I did not invite it into my intestines, share my food and even create many unwanted symptoms on me, makes me grow goose bumps. Perhaps it was my prejudice about these little parasites that prevented me from learning more about Jacob and Monod's work. However, Watson's book on DNA and E. Kandel's book in Search of Memory made good use of E. Coli secrets as revealed by those two researchers at the Pasteur Institute. E. Kandel came across a cycle response element-binding protein (CREB) for short. CREB is a gene regulatory protein involved in many processes including learning and memory. It binds to a promoter in a gene. In his laboratory he found that if he blocked the action of CREB in the nucleus of a sensory neuron, he prevented long-term, but not short term memory. It means to me that he was blocking signals to the central command center, the nucleus of the cell. Therefore, there were no orders for transcription to take place; consequently, there were no proteins in progression. Kandel wrote in his book: "Blocking this one regulatory protein blocked the entire process of long-term synaptic change!" The activation of CREB leads to the expression of genes that change the functions and the structure of cells. CREB proteins come, at least, in two forms, one for activation and another for suppression. This is the on- off switch nowadays. For us, it was not a surprise that protein synthesis takes place in the cell body. Oswald Steward from the University of California had shown us that protein synthesis also take place at synapses. It was quite confusing for us. It was confusing because it reverses the whole functional process of a neuron. It bypassed electro-genesis, hillock and the neuron nucleus itself. We will close E. Kandel in this chapter with the following: "The process initiates long-term synaptic facilitation by sending protein Kinase A to the nucleus to activate CREB, thereby turning on the effectors genes that encode the protein needed for the growth of new synaptic connections. The other process perpetuates memory storage by maintaining the newly grown synaptic terminals, a mechanism that requires local protein synthesis."[39]

---

[39]-E. Kandel, op cit, p, 270

I will add a final comment on CREB proteins and my companions, the micro RNAs. Among micro RNA there is a mi R 212 which is shaped like a hair pin. In experiments done with rats that were engineered to consume cocaine, the rats decreased their addiction when micro 212 increased. Perhaps this micro RNA can be translated into a therapeutic tool for human addiction. Further experiments demonstrated that blocking the formation of miR212 increased the appetite for the drug. At present time, we seem to have good command over RNA in its multiple forms. We can clone it, synthesize it and engineer it at will. This tiny RNA molecule volume is related to a CREB, a protein involved in long term memory and anatomical changes in neurons. Eric Kandel tells us that protein Kinase A goes to the nucleus of the neuron to activate CREB which in turn, turns on the gene that codifies the protein needed for growth of new synaptic connections. We beg you to bear with us our curiosity for ribonucleic acid. Whenever we turn a page in any book of biology and all its processes, there we find RNA involved in our lives. We focus our eyes and attention on tools that may help ameliorate the pain of our brothers and sisters worldwide. The pain outweighs the benefit, if any, when cocaine or heroin hijacks the brain of a person. "Protection against cocaine addiction may be a side benefit of miR normal job of regulating CREB production and other biochemical processes in the brain, said neuro-scientist, Paul Kenny."[40]

NOBEL LAUREATES ADVOCATING FOR RNA

A 1989 Nobel Laureate on chemistry for the discovery of RNA multiple functions, and in particular, ribozymes, wrote for Nobel Prize Org in 2004 that he, Thomas R. Cech and Sidney Altman, independently of each other, found out that RNA could fold into complex shapes. Furthermore, it catalyzes biochemical reactions, a function previously thought to be restricted to protein enzymes. Thus, RNA was sometimes an active participant in the chemistry of life, not just a passive messenger carrying molecules to a ribosome for protein synthesis. He added: RNA often controls the expression of genes, another role that had been thought to be at

[40]- Science News, Vol. 178, #3, 07/31/2010, p. 11

least mostly the domain of proteins called repressors and transcription factors." The discovery of the ribosome as a ribonucleic acid molecule that assembles protein just added scientific recognition of this acid vital role in chemistry, development and evolution of life on Earth. It was seen as a home-run for RNA advocates who were claiming that it has existed before the DNA world. RNA could replicate itself and multiply while DNA needs RNA molecule for the synthesis of life building blocks, proteins. One impressive feature of a ribosome is that it decodes a very large number of messenger RNA to make protein for each organ, tissue and function in our body. Can you imagine multiple ribosomes putting together all the units from a messenger and a transfer RNA molecule? Basically, all orders came out of the cell's nucleus for the purpose of building a protein for a specific organ in your body. For instance, the recipe or building formula for a specific cell in the pre-frontal cortex, heart, liver, lungs or a muscle in my left leg needs to be made in a precise fashion. There cannot be a mistake writing this formula. This is, by no means, a small job even for a summa cum laude graduate. Remember, ribosomes are not located within the membrane of the nucleus to be supervised like grandma used to. Ribosomes are located figuratively, miles away from the cell's nucleus, in the cytoplasm.

According to the above named researchers, a ribosome is composed of three RNA molecules (four in some species) and dozens of proteins. Most recent pictures of ribosomes at the atomic level show that it is in fact composed of RNA with no proteins in the vicinity. It is indisputable evidence that indeed it is an RNA catalyst. It is in charge directing mRNA and tRNA interactions in the process of protein assembly. At the same time that these discoveries were taking place, catalyst and enzymes were almost identical. Enzymes were the catalyst agent per excellence. In the past, RNA had been placed far behind, a passive agent you needed not to bother with. However, during the last decade of the XX century, and most recent discoveries, has dramatically changed our concept about this very old ribonucleic acid, its multiple shapes and functions. During the birth of the double helix in the early 1950s, there were hardly any books on RNA-if at all. They were hidden behind piles of illustrations of DNA and the double helix structure.

The third millennium brought us a new experience with our grandpa, the RNA molecule. We have seen how it turns gene expression on and off and plays a definite role in the making of proteins as well as chemical reactions within the cell. The beauty of this technological advancement in many areas of science is a blessing for humanity. The French and the American revolution of the XVIII century opened the door for our little genie that was kept imprisoned inside our skull to enjoy freedom. R. Descartes "cogito ergo sum" could no longer be kept a prisoner of the pineal gland deep in the brain. The French philosophers of late XVIII century knocked down many walls and the American brain built a new and democratic society. The human mind was set free. It could not any longer be kept enslaved in his body by an antiquated and obsolete body of knowledge. We challenged authority handed down to us by the force of the sword, prejudice and traditions that benefited and protected the few at the expense of the majority of humanity. Benjamin Franklin and Thomas Jefferson, although part of the American elite, their mind were not constrained by society's rule of behavior prevalent at the time. Franklin was a man of science ahead of his time. And on Jefferson, we are still debating his thoughts and actions. They were challenged by equally brilliant compatriots. We became the free men of planet Earth. Even the most repressive regimes used the word Democracy in reference to our system of government.

A pioneer neuro-scientist, Miguel L. Nicolelis, wrote that freeing the brain from the limits of our terrestrial bodies will allow the paralyzed to walk. Through this liberation of the human brain from the physical constrains imposed by the body, the disabled may rise from wheel-chairs. Humans have shown that a brain can be directly linked to machines in a laboratory setting. He adds: "A few years ago my group demonstrated the feasibility of linking living brain tissue to a variety of artificial tools. It will require a new generation of high density micro-electrodes that can be safely implanted in the human brain and provide reliable, long-term simultaneous recordings of electrical activity of thousands of neurons distributed across multiple brain locations." Doctor

Nicolelis envisions bi-directional, thought-driven interfaces, operating a myriad of nanotools that will serve as our new eyes, ears and hands. [41] Similar techniques like DBS or deep brain stimulation, is used today to assist Parkinson patients gain control of their body, among other things, controlling tremors and bodily balance.

Previously unknown RNA functions

Among its previous unknown functions are as follows: "It can provoke- cause- termination of transmission of messenger RNA, it can interfere with ribosomes from translating mRNA or it can cleavage (split or break up) mRNA into destruction. The ribonucleic acid molecule has a cleavage or splitting mechanism that seems to work even in itself. RNA existed at the side of DNA for most of the second half of the XX century. However, recently, with so many shapes and functions of RNA molecules discovered, it seems that, along with epigenetics, a new branch of molecular biology is needed. We are referring to RNA not only as a research tool, but primarily, as a medical drug treatment modality for many chronic and degenerative diseases that we are struggling to conquer. Our concern and interest is stopping the suffering of millions of people around the world. RNA molecules serving as a gene switch and protein most prominent assembly agent, in addition to breakdown facilitator, seems to have a very promising role in medicine in the very near future. Another attribute of RNA not previously suspected besides a formidable catalyst agent is its ability to come out as a double strand. Normally, RNA is a single strand, meaning that only one copy is transcribed from DNA. However, given that it is a single strand with no partner to link with, it possesses the intrinsic attribute of folding back into itself. It can take any shape necessary for a function. You can see RNA hair pin that are, functionally, double RNA strand. So, a single strand of RNA can fold or coil itself in different shapes. For the sake of clarification, you can see it as a long spaghetti strand, a hair pin or pair of pliers, a bead of a necklace or rosary beads or a sphere structure. It attaches itself to, or allows other molecules to

---

[41]- Scientific American, 02/2011, p.81-83.

link with as on histones in the chromatin. The linkage attribute takes many forms. In addition, you have the well-known RNA interference. This versatile molecule, RNA, can not only block protein synthesis, but with the assistance of enzymes like Dicer and Slicer, can cut double strand RNA into 20 base pair fragments. One of these two strands can be transported to a matching sequence of a messenger RNA for protein formation while the other strand may face degradation, meaning destruction. By now, you are probably almost convinced about the crucial role of RNA on your life. But this short story about RNA does not end in here. It is not just protein making and gene function interference. Its future use is waiting for you to take your place among our group of scientists.

And... talking about meddling on gene affairs, you have not forgotten RNA interference. It has become a very important technology for scientists to target gene inactivation in their never ending search for drugs for the treatment of genetic derived illness. Suppressing a gene may lead you to discover the causal factors of a disease or a syndrome leading to a disease in the brain or any other organ of your body. An RNA molecule that has the power to block the construction of the basic building parts of my body not only during the translation process, but also, inactivating my genes or your genes, is a powerful cluster of atoms. These atoms that are formed into a molecule we need to understand and gain to our side. Please, bear with us, because we consider your brain as extremely important for us to ignore just a tiny bit of it. Your protein factory, the ribosome consists of two subunits. It has a small subunit that decodes messenger RNA, a second larger subunit that serves as a catalyst for peptide bond formation between the growing polypeptide chain and each new amino acid. This is part of the process of translation we are committed to fully understand in eukaryote ribosomes which is much more complex and highly regulated than non-eukaryote ribosomes.[42]

A little bit of history

---

[42]-Science, 02/11/2011,Vol.331, p.681.

A little bit more of history of this ribonucleic acid does not hurt anyone. Basically, it began over a decade ago here in U.S.A., Europe and Japan. The race among scientists for the search of codons, amino acids, and protein synthesis got researchers on both sides of the Atlantic busy at work in their respective laboratories. The RNA molecule's many roles and shapes have motivated many biologists to decipher this molecular acid functions and shapes. The discoverers of the double helix in England were interested in this molecule that appeared to be engaged in all aspect of our life. Initially, we were happy reading and studying DNA double strand and RNA single strands such as messenger RNA and transference RNA involved in protein synthesis. However, as more functions were showing up, two young scientists: Craig Mello and Andy Fire had not overlooked the power of the RNA molecule to intervene in the process of protein formation. Both men, working on a little parasite that loves to live in mammal's intestines, including humans, E. Coli, provided the first ever demonstration that double stranded RNA could interfere with genes and even silence it. Do not forget this statement- it is very important. Once again, the RNA molecule that is so important during the synthesis of proteins, can intervene during the whole process of making proteins for your body and mine. Among other things, it can bind to a messenger RNA that comes from the DNA nucleus and derail its synthesis. In addition, please, do not be alarmed, it can also silence genes. Mello and Fire posed a huge challenge to biologists around the world. To think that a ribonucleic acid molecule can alter and even stop the transcription process, a dominant function of DNA molecule, was heretical. Ever since I first read an article on this interfering molecule, I began to wonder what did RNA have to gain when it decided to become DNA's molecule workhorse. Within each cell we have three wise and powerful molecules working together to build, develop and maintain a very complex organism like yours and mine. The three king molecules we are referring to are: DNA, RNA and mitochondria. Each one has definite and clear roles to play and they have played it very wisely and beautifully. When I consider that silencing RNA molecules can be used to mark specific DNA sequences for deletion, and guide modification of chromatin structure, I get goose bumps. However, I try to bring relief to myself by saying:

This molecule has worked remarkably well on me. The mitochondria with its ATP molecule cannot be counted out during the balance of power among the three rulers of our chemical factory, our cell. A recently published scientific journal (2011), a team of researchers who made use of mice mitochondria DNA linked to health benefits. In mice, a regime of endurance training-running for 45 minutes three times a week over five months induced mitochondria bio-genesis, increased mitochondria respiratory capacity, and prevented mitochondria damage."[43]

An established fact, an RNA world

An RNA world existing before a protein coding one, which is the DNA world, is an established fact among modern scientists. The irony concerning this scientific truth is that it became accepted after the discovery of the double helix. Ironically, it provoked us into laughter and further self-questioning about its functions. The seemingly paradox that we are trying to deal with is that multiple cell complex organism developed after the protein coding DNA world took over. However, the work horse in protein synthesis is the RNA molecule in its multiple forms. We have said that the DNA molecule incorporated or swallowed in RNA to do its job in the process of forming or constructing proteins. Similarly, we also pointed out that the wise DNA molecule incorporated the mitochondria. We have not been able to resolve this issue to our satisfaction. However, we are very happy with our progress understanding both macro-molecules and the sequencing of the human genome. We would like to share with you an excellent article written by Eric S. Lander in a scientific journal. We will write down just a few things, although incomplete, but it sounded very interesting to us. The sequence of the human genome has dramatically accelerated biomedical research... and our understanding of the biological functions encoded in the genome. The greatest impact of genomic has been the ability to investigate biological phenomena in a comprehensive unbiased, hypothesis free manner. In basic biology, it has reshaped our view of genome physiology, including the role of protein-coding genes, non-coding

---

[43]-Science, Vol. 331, 03/04/ 2011, p. 1115.

RNA and regulatory sequences. In medicine, genomics has provided the first systematic approach to discover the genes and cellular pathways underlying disease. He further added the discoveries of locations where DNA and chromatin are modified, in addition to the study of inherited variations or somatic mutations. Sequencing is also applied to RNA transcripts and its multiple forms. We have also clarified the volume of protein-coding genes in the human genome as consisting of about 21,000 genes. At the beginning of the genome sequence project, some researchers had put the number of protein-coding genes as high as 150,000. On small non-coding RNA, Eric S. Lander said that a few dozen mi RNA has been shown to have key regulatory roles. Most recently, a new class of small RNA called Piwi interacting RNAs, has been discovered; they act to silence transposons in the germline. Transposons may be seen as drivers of evolutionary innovations. In addition, we have benefited from sequencing the genome, the identification, location and roles played by epigenetic markers. The vast majority of human variants have been discovered, and they are under intense analysis. In its clinical application, DNA sequencing is being increasingly used to assign patients with an unclear diagnosis to a known disease. In psychiatric diseases, genomic studies have identified common variants in bipolar disorder and schizophrenia and rare deletions in autism.[44] We could not provide a complete account of the article, but we hope you can get access to the journal and enjoy it as we do.

Dr.Tom Misteli from the NCI and DNA

Ten years ago, a blueprint for a human being, the complete list of the DNA now famous letters ATCG was announced to the world. In celebrating a decade of genome research, Tom Misteli, a senior investigator from the National Cancer Institute in Bethesda, Md. said: "Biologist have known for long that DNA chromosome folds up in complex ways. They have now demonstrated that individual chromosomes occupy distinct territories in the nucleus and that some chromosomes prefer the nucleus periphery, whereas

---

[44]-Nature, Vol. 470, #7333, 02/10/2011, p.187-197.

others like to cluster close to the core. Moreover, where chromosome resides, and which chromosome lie near one another, can strongly influence how cells function." (We did not fail to notice the word –strongly- to define the influence on the cell.) Tom further added: Where chromosomes lives seem to influence whether the genes it carries are turned on or off."[45] A gene gets turned on after proteins known to us as transcription factors get together on regulatory regions of the gene. Our workhorse, the RNA, this time in the form of protein RNA polymerase, comes in to transcribe the gene's DNA famous four letters. They are transcribed into RNA multiple copies you are already familiar with. DNA you will find it in the nucleus, while RNA is always moving around and changing its shapes and functions. Following Professor Misteli's observations, he generalized and said that researchers now know that the nuclear periphery has silencing effects on genes, while the center promotes activation. He further elucidates the intrinsic working processes taking place during protein synthesis. He added that hundreds of genes that encode ribosomal RNAs are transcribed together in the nucleolus- a nuclear substructure large enough to see under a microscope. [46] The nucleolus resides within the nucleus, but has its own membrane. Do not look for it around ribosomes or nearby organelles. Within the cell there is a good busy traffic going on from one area of the cell to another. You will find organelles transporting subunits from the cell's basic components, from one station within the cell to another. You will also see polymers brought into the cell through its membrane as well as sharing or transporting material outside to neighboring cells. If your organism has been invaded by unwelcomed bacteria and viruses, please, use your imagination for the traffic jam that will take place in millions upon millions of cells in your body. Your T and B cells, with legions of combatants, will order an attack to eliminate toxic invaders.

Micro RNA regulatory role

[45] - Scientific America, 01/ 2011,p. 70-73
[46] - Scientific American, 02/2011, p.71

The celebration a decade of genome sequencing cannot ignore micro-RNA's regulatory role in gene expression. "Over the past decade, small RNA emerged as a new class of key regulators of eukaryotic biology. This diverse class of RNAs includes small interfering RNA, micro-RNA and PIWI interacting RNA, all of which associate with multiple protein components within a complex to regulate partially or perfectly complementary transcripts. Among the objectives the researchers were trying to understand is how mi RNAs bound to Argonate proteins, bringing about gene silencing. There is abundant literature indicating that translational inhibition and m RNA decay are coupled throughout biology. [47]

Unbelievably naïve

Our intention during this essay has been somewhat naïve. Our intention was to analyze and understand the code of life Watson and Crick began to play with during the middle of the XX century. We began with the double helix and proceeded to nucleotides, amino acids and protein formation which are among the most interesting stops we found in our tree of life. We stopped for a while at the intriguing RNA molecule and its many curious and useful conformations and functions it takes to serve us best. Of course, we could not bypass epigenetics and multiple DNA mutations and histones modifications which were briefly touched on. Perhaps my intention during this trip, although unconscious, is the very simple and naïve question of my origin as I am now. The nagging unknown of my original birth on planet Earth millions of years ago. I will not be able to solve it, but you will. My own ancestors, including my grandparents, did not have this mind bugging urge to explain the unknown. They relied upon generations of old tales often written in their sacred book, the Bible. More recently, our origin has been attributed to aliens or meteorites that came from another solar system. Both, my ancestors and recent theorists, are not too far apart; both place my origin outside planet Earth. However, for some reason, we tend to stick to the "chemical soup" theory of mother Earth. This wonder

---

[47] - Science, Vol.331, 02/04/2011, p. 550-53.

soup, under certain pressure and heat, attracted selective atoms and molecules that in due time became living cells. It has been a long, long time ago in our Earth's history that this spark of life took place, if at all. Most recently, science fiction and some respectable physicists and astronomers are advancing a theory that aliens with super- intelligence may have visited Earth earlier in time. We are just guinea pigs under constant scrutiny and observation. It looks as if we are animals under a laboratory's watch to see how we behave under certain conditions. How we take care of ourselves, and ultimately, take care of planet Earth. This one might be a feasible goal.

Here and now with amino acids and proteins

All the above theories sound very interesting, but we prefer to stay close to mother Earth. Instead of traveling out of our solar system in search of myself, I am stuck with amino acids, peptides and Earth elements that formed my life here. Of course, the double helix and the gene code is a giant step forward that keeps my attention 24/7.

But, when did life begin in this beautiful baby? Mom rightfully would say, "This is my baby, I gave birth to him." Well, you pass judgment over mom's claim. Did life begin at conception? Or should we say during the first cell division. Perhaps we can wait three days later during the morula phase of embryonic development with an internal mass of eight cells? Someone might argue that we must wait another three more days until the phase of blastocyst takes place and cells begin their journey to specification. Perhaps, a wiser man joined the conversation and argued: "Life does not begin until cell specialization has begun with cells for a heart, brain, lungs, liver and the remaining organs and tissues of our body are in place." You are right, we are not now talking about the same thing. The original unknown was left far behind, but this is what is happening during most of the meetings I have attended to on this subject. My inquisitor may interrupt me and say. Oh, but this is democracy, my friend! Of course, it is democracy, but we are not talking about democracy. Besides, democracy in itself is in need of a new definition.

St. Agustin and Darwin join our group

Suppose that our group of science curious individuals is joined by someone like Socrates arguing in favor of his soul? Even a better scenario would be Aristotle, St. Agustin, Darwin, Einstein and Oppenheimer joining our lively group. It seems to me that in this hypothetical case, we have a centrifugal chemical soup that, nevertheless, could result in an implosion provoking a tentative truth. But, sadly to say, whatever truth the implosion group may come out with, it is an educated guess of a group of people in a bit of dust in an infinite multiverse. Please, do not blame me for beginning this mind torturing dialogue. I blame the people I named above including Watson and Crick. In the meantime, we will continue our journey without more outside interruption. In humans, brain growth and differentiation comes under the control of genes around five to six weeks after conception. However, this pre-natal phase of development is very critical for the fetus's health. His mother's diet, life style and environmental factors will influence the fetus development long after it leaves the comfort and protection of the mother's womb. We are all aware of cigarette smoking, drug abuse, biological predisposition, hunger, physical abuse and chronic stress during pregnancy.

An estimated 100 billion neurons and around 800 billion glia cells need the correct amount of vitamins and minerals to make their appropriate and corresponding shape. They have to travel within the cell building and discharging neurotransmitters necessary for final normal function. A deficiency in the basic elements necessary to build amino acids that ultimately will form proteins for our body is well established around the world's scientist community. Following each step during the fetus development is not the best thing to do with our present technology. We have our four legged friend, the mouse, and other animals in our laboratories at our comfort and convenience. We use them to mimic human like dietary deficits and possible health risks.

Stress and ADHD in pregnant mothers

Let us take stress as a subject of study. In a pregnant mouse, the exposure of the future mother's stress hormones can lead to anxious behavior and hyperactivity in the offspring. A relatively recent longitudinal study of over 7,000 mothers and babies (humans, not mice) at Imperial College in London concluded that maternal stress may account for up to 15 percent of diagnoses of attention deficit hyperactive disorder. [48] During the first decade of life, boys and girls spend much of their time playing together often under loving parental supervision and control. During this period of human development the central nervous system is busy making many necessary circuits and pathways. During adolescent hood, sexual hormones are having a big party inside your body. In the meantime, glia cells and neuron's dendrites and spines are trimmed to serve specific purposes corresponding to biological, physiological and behavioral stages of development going on during this critical period of human growth. You have association neurons, motor neurons, visual, acoustic, limbic system, and many other brain cells making appropriate adjustment to best serve our organism.

The pre-frontal cortex, the brain area par excellence for cognition and high level decision making is the last neocortex area or region to complete maturation. Unfortunately, this is the time when adolescents feel that they have to experiment with everything around them that adult people take for granted. We experiment with drinking alcohol, smoking and car driving. Car accidents resulting in injuries and death are higher during this period of our life time span than any other period of time. Unfortunately, it takes place while our brain's decision making cluster of cells is still in the process of making corresponding axonal connections. It does not mean that connections are permanently closed. Synapses are busy receiving and sending out chemical messages with potential new pathways. Adding to adolescent hood turmoil, a dreadful brain disease, schizophrenia, takes its toll during late teens and early twenties. Brain areas like nucleus accumbens and ventral

---

48- New Scientist, 04/04/2009, p.28

tegmental area are easily recruited by our reward system which ends up hijacking the whole brain. Consequently, addiction takes over the person's life. According to reliable studies in different laboratories, once an addiction drug has hijacked the brain, it provokes permanent changes to those areas. The addicted individual exists to sustain an addiction that has taken over part of his brain that dictates his behavior at all cost. The addicted individual does not enjoy his drug because the state of feeling that you and I experience, is no longer part of their brain behavior.

Day dreaming

Our hypothetical group in which St. Agustin and Darwin had joined us to help to come to an educated guess is based on many TV documentaries that have attempted to solve our problems. The groups have ended, in the majority of cases, just like in the beginning, each member holding to his or her beliefs. So, we were not dreaming, we were, I guess, mind traveling for a short time. But, is day dreaming a resting period of the brain? Day dreaming has had multiple interpretations and meanings for many people around the world. Some researchers have concluded that it allows brain memories to be analyzed and organized into content related compartments in brain cells. Occasionally, researchers and psychologists have associated daydreaming with neuron affinity, circuits and pathways compatibility and normal functioning. Much less reliable are philosophical and religiously inclined individuals who postulate that daydreaming is the time for the soul to leave the body momentarily. They maintain that the soul leaves the body to communicate with its superiors, whatever it might be. All the above persons defend each other's position with about equal tenacity and conviction. Unsurprised, when I began to read an article on day dreaming that related to brain diseases like schizophrenia and autism, my brain was half asleep. Occasionally, and for some individuals, busy day dreaming suddenly lightens up the brain like the sun at noon time. Day dreaming can be a very sweet thing, especially when boys dream with a girl that exists in dreams only. Excessive day dreaming can be an escape from reality and challenges of everyday life demands. But poets, story

105

tellers and novel writers claim their best literary work was done during day dreaming.

The brain, just 2 percent of your body mass, but...

The brain, the day dreaming part of your body, is only two percent of your body mass (not always true if you are an obese person); but that brain of yours consumes twenty percent of the calories you and I eat. I could not help it, but my mind jumped immediately to implicate the replication, transcription and translation processes RNA is involved with to form proteins in my body. The mitochondria and its energy producing molecule, ATP, popped up in my thinking processes and alerted me not to count them out. The truth is that the resting brain, as it is during sleep, especially REM sleep, is not resting at all. REM stands for rapid eye movement. During my REM sleep, in all probability, my brain is consuming more energy than while I write these lines. I repeat it over again. The brain with only 2 percent of the body mass, when it comes to calories consumption, it seems a vacuum cleaner that gobbles up 1/5 of all converted calories you have sent down your throat. Even more frightening yet, your brain does not store energy for leaner times. If calories are needed and there is no food for you to bite or swallow a drink and throw it down your digestive pipe, fat that has been stored in your belly, buttocks, thighs, among others, will be going to feed your brain. If for some reason you contemplate and risk going on strike against your brain's demand for calories, you are on the losing side of the battle. Your brain has a mechanism that will convert whatever you have stored in your body into calories until there is no more to devour and death arrives.

The scenario we presented you with a brain's need for a huge amount of calories is not a hypothetical case. A group of researchers whose curiosity in that brain of yours need for calories, wanted to know why and how does your brain consume so much energy. It consumes energy even during the resting state of brain functioning. To tackle this curiosity issue, they have used the most modern technology at their disposal, PET and MRI scanners. During one experiment, brain cells were fully activated during rest time, but quiet- seemingly resting- as soon as the person began to mentally work on an exercise. Seemingly paradoxically, but a very interesting experiment, it was. During this experiment, it seems that if I work on solving a puzzle, my brain cells leave me alone and go on its own business. However, if I go into rest stage, the brain neurons go to work. My brain seems to have a funny work habit.

Raichle, Shulman and the default network mode

Gordon Shulman, a researcher, went through 134 brain scans and found that regardless whether the subjects in the experiment were reading or watching shapes on a screen, the same group of brain cells began to darken out. They were getting dim as soon as they began mental concentration. The researcher called it the default mode. Does it mean that brain cells chat among themselves when not busy doing routine day life chorus and complex intellectual work? This is even more curious. During the resting or chatting time, parts of the network consumed 30 percent more calories, gram for gram, than nearly any other part of the brain. [49] Well, you can argue that the brain's multiple neuronal processes including electro- genesis, memory formation and storage as well as energy supervising autonomous nervous system, demand caloric energy in vast amounts. Our brain is always busy in mind traveling, solving routine daily life chores, planning self-protection or trying to understand Einstein's quantum theory. That brain of yours deserves a lot of respect. It led you out of the jungle and is planning to take you out of planet Earth and go where no man has gone before. These researchers used very modern tools to come to very interesting comments and conclusions we cannot avoid sharing with you. Marcus Raichle says that there is a huge amount of activity in the resting brain unaccounted for. Making use of PET scanning, he noticed some brain areas seemed to get full tilt during rest, but quietened down as soon as the person exercised. Neurons chattered non-stop to one another when an individual is unoccupied, but as soon as a task requiring focused attention from a person comes in, chatting stops or quiets down. We became even more interested on this discovery when we learned that the median

---

[49] - New Scientist, Nov. 8-14, 2008, p. 29.

pre-frontal cortex and the hippocampus become involved in this default network. You already know that the hippocampus is involved in the formation and retrieval of memories. The pre-frontal cortex is highly engaged in decision making after receiving input from many important brain areas. A possible task for the default network would be linking multiple neuronal circuits and pathways among clusters of brain cells and provide the brain with an inner rehearsal for considering future actions and choices. Using fMRI, Malia Mason of Dartmouth College in New Hampshire, equated this network mode with day dreaming. Randy Buckner and Daniel Gilbert from Harvard University "see it as the ultimate tool for incorporating lessons learned in the past into our plans for the future." Marcus Raichle now believes that the default network is involved, selectively storing and updating memories based on their importance from a personal perspective. He added that rather than burning the extra glucose for energy, it uses it as a raw material for making the amino acids and neurotransmitters it needs to build and maintain synapses. [50]

Dr. Marcus Raichle's opinion

However, another scientist using MRI reported that day dreaming took place when the default network was active. Default mode was coined for an unoccupied brain. The brain is not working on your own business. It must mean that you are consciously working on something. Marcus Raichle from Washington University at St. Louis believes that the default network is involved selectively storing and updating memories. Another discovery of Reichle on our brain is that it has a sweet tooth. It devours huge amount of glucose, way out of proportion of oxygen it uses. He believes that the extra glucose the brain consumes is use as raw material for the making of amino acids and neurotransmitters that the brain needs to function properly. Now comes the big surprise: Raichle and two more researchers found that the default network's pattern activity is disrupted in patients with Alzheimer brain disease. This author adds that most Alzheimer patient's brain is not occupied with anything because

---

[50] - New Scientist, Nov. 8- 14, 2006, p. 31.

their memories are gone and their neurons are blocked by plaques. And... organelles inside their neurons are entangled to one another, and each one is unable to send or receive messages.

The point that I am making is that chatting among brain cells in people suffering from advanced Alzheimer disease is very questionable. However, Raichle's findings indicated that the default network also turns out to be disrupted in other brain maladies including depression, attention deficit hyperactive disorder, autism and schizophrenia. [51] Neuro-scientist Elkhonon Goldberg, from New York University was commenting on the frontal lobe and Alzheimer disease and wrote. "My collegians and I have shown that frontal lobes become dysfunctional at a very early stage of Alzheimer's type dementia...The frontal lobes are more vulnerable and are affected in a broader range of brain disorders: neuro-developmental, neuro-psychiatric, and so on, than any other part of the brain."[52] The findings in this article made me scratch my left ear because the brain in schizophrenic patients, A.D.H.D., and autism in a lesser degree, are on high gear most of the time. Patients diagnosed with major depression, their brain activity as shown in scanners seems to be in siesta time most of the time. The way we see it, brain neurons in schizophrenics and attention deficit hyperactive disorders would be chatting most of the time forming hallucinations and delusions. On the other hand, we have the deeply depressed person who complains that she or he has nothing to report during psycho-therapeutic sessions. Many depressed patients complain that his or her brain is empty. I recall them saying to me, 'my head is like an empty shell, nothing goes in and nothing come out.' Actually, medication prescribed to them is intended to activate their neurons.

Pete, my autistic boy in my office

Our experience with autistic children has been limited, but we observed that they enjoyed themselves when I gave Pete a game or puzzle to work on. I sat quietly next to him. I remember Pete telling me the kind of puzzle he used to love to work on. He

---

[51]- New Scientist, 11/08/ 2008, p. 31
[52] - The Executive Brain, Elkhonon, Goldberg, Oxford University Press, 2001, p. 115

began trusting me, taking little things home from my office. However, he returned it to me the following session. He strongly disliked noise and excessive sunlight. He used to pull down the shades in my office. Another interesting observation I remember from his behavior is that he did not like his brother or his mother to share our sessions. During most of the time we spent together, he began our conversation. At the end of each session, Pete often asked me for an object to take home. I liked him asking for something to take home from me. It had significant meaning for me. It took me time to refer him to a licensed trained therapist. I had neither the training nor the neurological understanding underlying autistic behavior. We can interpret him taking home toys from my office as taking home part of me, thus bending isolation through trust. How to use brain plasticity to normalize brain circuits and chemistry was the key question that I was always trying to deal with while in practice. With Pete, I confess I was frustrated and sad I did not know how to help him.

Autistic children are hypersensitive to sound and touch. For them, a soothing bird song, sounds like a drill sergeant during basic training, or an 18 wheeler truck blowing its horn while I drive my four piston small car in the highway. Most recent scan studies have revealed that autistic children process sound in an abnormal fashion. Likewise, a soft skin touch to an autistic child seems to be the rough game of soccer or football when they jump one on top of another. "Autism is largely an inherited condition. If one identical twin is autistic, there is an 80 to 90 percent chance the other twin will be autistic as well." [53]Autistic children do not seem to be able to discriminate, consequently, tolerate, degrees of stimuli. It seems to us that the autistic brain closed up neuronal circuit and pathways in selective brain areas ahead of time before it could establish stimuli control levels. They are unable to decrease the sound force of a thunder in the sky and bring it down to tolerance level. Sound frequencies have to be established. Special training and teaching in general will help to re-educate neuronal circuits and pathways leading to normal responses. Autism is much more prevalent in boys than girls. The autistic child is not avoiding

---

[53]- The Brain that changes itself, Norman Doidge, Penguin Books, N.Y. 2007, p.77.

or rejecting anyone. He or she is protecting himself or herself from experienced harmful stimuli. The earliest we establish the diagnosis, the sooner we can begin synaptic re-education.

Isolation and loneliness in Schizos and autistic individuals

Schizophrenic and autistic individuals are socially isolated people. Their brain seems to be on the alert tone most of the time. It seems to be as if they are expecting rejection from all sources. Consequently, they respond by escaping into their own shell which is comforting and safe. Researchers have found that the amygdala as well as the pre-frontal cortex is more active than most of other brain regions. We may also interpret the hyperactive amygdala and pre-frontal cortex as an adaptation or adjustment of those areas. It appears to be autistic children's behavioral respond towards people they come in contact with. There may be a biological pre-disposition that is triggered or activated by outside behavior. Their brain is not engage in regular daily problems and conversation, consequently, they turn inward. We cannot call them day dreamers because the brain is very active forming or recalling hallucinations and delusions or forming respond designs. You will find similar brain behavior in bipolar- manic phase individuals. The pre-frontal cortex is actively judging rejection signals, therefore, common sense behavior, decision making and academic accomplishment become deficient not because they are mentally retarded, but because their brain is on the super- alert switch. Consequently, focusing attention is a major problem. This may lead to feeling of chronic sadness; not that sadness is inherited, but along with loneliness, is an undesirable behavior coming from a biological predisposition. The environment, particularly, during the early years of development is very critical. We must remember that the brain is not a rigidly wired organ of our body. We have new neurons forming every day, particularly, in the hippocampus and ventricles. In addition, we have synapses connecting neurons and establishing new pathways reinforcing good and appropriate behavior and thinking. Brain plasticity is for us to take advantage of and put to work with people in need. The brain learns new things every day. It learns something new and the region or cluster of neurons responsible for processing sound and touch in autistic children have to adapt to new levels of signals.

Neurobiologist Jeff Lichtman from Harvard University hopes that studying the structural changes in the brain will eventually show him how information is encoded in our brain wiring. Using a method he dubbed Brainbow, a small piece of DNA with genes that code for random amounts of yellow, blue and red fluorescent proteins in nerve cells are inserted into mice. Through this method or research tool, his team can give individual neurons different colors. Those colors will allow researchers the opportunity to follow and determine synapse and axon or axons connectivity. Connections offer a way to investigate thought disorders, such as mental illness and learning disabilities. Further addressing the wiring or connections between neurons, he says: The plasticity of the brain's wiring is one of its greatest strength and scientists hoping to learn more about how we learn and adapt as we grow."[54]

YOUR BRAIN AND PLASTICITY

Aristotle, the ancient Greek philosopher and tutor to Alexander the Great, proclaimed some truth on personal observations. They became absolute dogmas for almost two thousand years. Accordingly, the sun and stars were moving around planet Earth as if it were the center of the universe. Our dear and sweet mother Earth was believed to be a flat mass of land until a few hundred years ago when Christopher Columbus helped tear down that fallacy. People become dogmatic and stick to beliefs and practice even when facts contradict it. And, we should point out that it happens not only to the uneducated masses. The theory of localization as pronounced by Paul Broca and Carl Wernicke became an axiom, an unchallenged truth still persisting in some professional quarters. Dr. Broca had a patient that answered all his questions with the word, Tan, tan, after suffering from a stroke in 1861. After the patient's death, he performed brain surgery and found out that an area in the left hemisphere, now known as Broca area, was gone. A few years later, Dr. Carl

---

[54]- Science Illustrated, March-April, 2010, p.35.

Wernicke had a patient with a similar problem, except that this time, Wernicke's patient spoke in an unintelligible fashion. It sounded more like gibberish than a normal conversation. It lacked conjunctions among words and phrases. After surgery, a region in the left upper temporal lobe, now known as Wernicke's area, was caput; it was damaged. Along these two cases, followed by many well attended professional conferences, the one function - one location theory, was established. Both areas, Broca and Wernicke, are located in the left hemisphere of the brain, thus, another "truth" was established and the left hemisphere is dominant for language.

A non-dogma believer, J. Cotard

However, a non-dogma believer and science oriented physician, Jules Cotard, studied children who had early massive disease during the year of 1868. He found among those children that the left hemisphere, including Boca's area, was wasted away. Yet, these children could speak normally. In 1876, Otto Soltman removed the motor cortex from infant dogs and rabbits. The motor cortex is responsible for movement. We will address this shortly. Another surprise, Soltman found out that his little rabbits and dogs could move. Both, Cotard and Soltman had not only disproved the dogma one function, one location, but more importantly, they had discovered the plasticity of the brain. Both experiments had shown that the brain was plastic enough to re-organize itself to meet demands of daily life chores. From Broca's surgery in 1861 until now, 150 years have passed and we still have many professionals doubting the brain's capacity to take over functions damaged by strokes or a disease. Furthermore, despite massive evidence using the most modern scanners, it is very hard for some individuals to accept the growth of new neurons in the brain. Plasticity of the brain must be understood not only taking place at synaptic level, but clusters of neurons taking over functions from one brain hemisphere to another. Revisiting brain neuro-genesis- a recent study (10/2011) reported that the subventrical zone around the anterior lateral ventricles in the infant human is highly active, producing many tangential migrating immature neurons. However, beyond 18 months of age, both proliferative activity and cells expressing markers of immature neurons are largely depleted.[55] Arellano and Rakic comenting on the above quotation wrote that Sanai et al. show that the migratory cell streams departing from the SVZ are present in human newborns, but disappear by the 18th month after birth.[56]

The year 1876 in medicine is a long time ago.

In reference to Otto Stoltman's work with kittens and puppies motor cortex, we do not know what he did. It is a complex system composed of multiple regions in the brain. His surgery was performed on infant animals when the wiring of their brains was taking place or in the process of consolidating functions. What areas or groups of motor neurons he removed is unknown to us. It was 1876 and not a single machine we have today to slice and examine brain organs, tissues and functions existed at the time. We decided to write a few observations on this subject so you can appreciate the strain efforts a Parkinson affected person goes through to be able to get up, move and walk. Similarly, we have family members, friends and valiant soldiers who come home with spinal and brain injuries. The motor cortex has multiple loops facilitating lateral horizontal, ascending and descending connections with the neo-cortex and spinal cord, among others. It is divided into three reciprocally interconnected areas, the primary cortex, supplementary motor area and pre-motor area. In the primary motor area, neurons correlate with a variety of movements and firing of individual cells. The secondary motor cortex is involved in planning movements and the pre-motor area is particularly concerned with planning movements that require sensory cues. [57] With Parkinson patients and combat soldiers with central nervous system injuries, it is very painful to overcome the limitations imposed by these brain deficits. It is not only the physical, but the psychological stress coupled with the injury. These are adult men and women who have to learn anew how to get up, move and walk. They have to force neurons in different brain locations to learn to do a job they had not done before. It is a

---

[55] - Nature, 10/20/2011, p.382-385.
[56] - Nature, 10/20/2011, p.333
[57] - Neuroscience, A. Longstaff, Bios Publications, U.K, 2000, p.221-22.

physiological and psychological demand on the patient to force plasticity of brain cells to take a new function. It is a feeling I cannot share with you in words when I see patients, family members and friends walking and doing things they could not do before.

P. Bach –y- Rita and plasticity

In 1969, Europe's most prestigious scientific journal, Nature, published an article that made many people shake their heads in disbelief. The article, whose leading author was scientist and physician, Paul Bach-y-Rita described a device that enabled people who had been blind from birth to see. My initial response would be that he was talking about science fiction or the biblical story of Jesus of Nazareth two thousand years ago; however, it has been proven that it was neither. Almost half a century ago, Doctor Rita had rejected the above dogma of one function, one location and proclaimed that our senses have an unexpectedly plastic nature. Rita maintained that if one location is damaged, another can sometimes take over for it, a process he called sensory substitution. He claims we see with our brain, not with our eyes. It reminds me of Geordi La Forge in Captain Piccard's starship Enterprise. Dr. Rita developed ways of triggering sensory substitution and devices that gives us super senses. He laid the groundwork for the greatest hope for the blind: retinal implants, which can be surgically inserted into the eye.[58]

Self-isolation, not loneliness

Self-isolation was widely practiced by monks during early years of Christianity. Socrates' dual nature of humanity, body and soul, was well accepted by Christians for the purpose of restraining bodily imperfections and desires. Consequently, it advances the soul's journey while on planet Earth. Today, we advance the needs of the body and the thinking processes of our brain. The soul has

---

[58]-The Brain that Changes itself, Norman Doidge, Penguin Books, N.Y, 2007. pgs. 10-17.

been left out to follow its own destiny. Self-imposed isolation, except during hibernation in some animals, is not our preferred way of life. However, our curiosity for all living things stopped at a bug buried in ice for 120,000 years. Do not hesitate, there is nothing wrong with your eyes, it means one hundred and twenty thousand years. Scientist coaxed it back to life! The bug was dormant in three kilometers deep in the Greenland ice sheet. We asked ourselves, what did it eat to survive so many years under such an inhospitable environment? The research team leader, Jennifer L. Curtze named dust, bacterial cells, fungal spores, minerals and other organic debris as a source for food. She further added that the best medium to preserve amino acids, organic compounds and cells is ice.[59] We like to add that Jupiter's moon, Europa, and both poles on planet Mars may well be home for primitive life forms to thrive on. In 2010, scientists from NASA reported that a bacterium in a lake somewhere in California could survive in arsenic acid which is fatal for all of us. This is not science fiction or a matter of faith; these are scientific facts. The journal, New Scientist, dated 10/18/2008, on page 17 reported: DNA of gold- mine bug may be key to alien life. The article claimed that "the organism's ability to live in complete isolation from other species, or even light or oxygen, suggest it could be the key to life on other planets." This strange bug was found in a South African gold mine nearly three kilometers beneath Earth's surface. Dylan Chivian, from Lawrence Berkeley National Laboratory in California, analyzed the newly found bug and said that the bacterium gets its energy from radio active decay of uranium in the surrounding rocks. It has genes to extract carbon and nitrogen from the environment- both essential for protein synthesis."

Evolution, adaptation and surprises of all sorts continue to keep my brain in a permanent state of alert for new discoveries. In an October 2007 edition of New Scientist, I read that there is a possum that slept for a year. The researcher claims that a possum he was experimenting with, after stuffing itself with food, curled up and went to sleep in his laboratory for a full 367 days. Another animal named, Zapus princeps, had slept for 320 days, however, an

---

[59]-New Scientist) 6/20/ 2009, p. 8.

Australian pygmy possum broke record by using one-fortieth of the energy it does while awake.

The hunt for diseases and the genome sequencing

The hunt for diseases in genes has engaged many outstanding scientists around the world. Initially, the main focus of attention was centralized in DNA's letter changes or mutations that were suspected of fatal diseases such as multiple sclerosis, Alzheimer dementia, schizophrenia and autism, among others. In some of these cases, the gene may appear normal, but there might have been a surplus or deficit of DNA sequences. Among copy number variations, we came across chromosome 21 with three copies instead of the normal pair, consequently, provoking the Down syndrome. Past literature refers to this syndrome as Mongolian looking children. This type of rare variation was considered conducive to diseases in most case. However, further studies revealed that variations in genes quantity is not an anomaly, but on the contrary, is quite common. In the year 2006, a group of geneticist analyzed DNA from 270 individuals and identified an average of 47 copy number variants per person. The genome sequence of C. Venter, a pioneer in the field, was found to have 62 copy number variants. Already mentioned in another chapter, J. D. Watson's genome also revealed a good number of copy number variants. These two outstanding scientists are doing exceptionally good work for humanity; so you cannot, but wonder what is the role or function of those rare variants. However, "copy number variants have been linked to disease like autism and schizophrenia."[60] We can argue that Down syndrome rare variants should not be placed along autism and schizophrenia, and we fully agree with it. We have made the observation several times. These last two diseases as well as bipolar, we consider them syndromes. It is not only a matter of genes, proteins, cells signaling, neurotransmitters and glia cells, but a myriad of other causal factors. By the way, copy number variations have been described as alterations of the DNA of a genome. It provokes the cell to have an abnormal number of copies of one or more segments of the

---

[60]- Scientific American, 06/ 2009, p. 24-25.

117

DNA molecule. The alteration may be an increase or deletion of a segment of the genome. In some cases, the variations may each be as much as 10 percent of the human genome. It may pose possible serious risk factors for diseases or abnormalities nobody wants to have. The scientific opinion on the etiology of these genome variations seems to be divided at present time. There is evidence that it is inherited while there is scientific support for de novo mutation, meaning that it may be seen as spontaneous mutation.

Mobility within the genome

We must have clear in our brains that the genome is not a long strand of beads tied to each other without mobility. The genome suffers deletions, modifications as in SNPs, duplications as well as multiple interventions by genome fragments. There are several gene segments that are involved in DNA-RNA multiple processes, consequently, variations in general seem to be the rule instead of the exemption. Copy number may take place in a single gene or be extended to a neighboring set of genes. The quantity factor, meaning too many or too few in relation to risk propensity, genes may be responsible for a good substantial amount of human phenotype variability. Besides, we have to consider behavioral traits and disease provoking sensitivity secondary to rare variants.

James Lupski, a clinical geneticist at Baylor College of Medicine in Houston, Texas, published online (10/14/2010) in the prestigious scientific journal, Nature, "highlighting the importance of rare genetic variants in causing diseases...even one copy of a certain gene can have profound consequences for brain development and mental disabilities." We add, How about gene mutations that do not code for proteins and become degenerated debris in a cell? The mitochondria DNA is also subject to multiple modifications and problems. This female contribution to our body and heritage can be a blessing or a liability. Dominant and recessive variants on both sides can contribute to make me a genius or a disabled individual unable to care for myself. One of my parents gave me an unwanted present, thalassemia minor. Purification of our genome has to be kept on the table for future laboratory consideration and analysis.

Reinforcing erroneous assumptions

Not long ago, many scientists used to believe that silent mutations were inconsequential to health because those changes in DNA would not interfere with protein synthesis. Although some disorders were traced back to a silent mutation, the general consensus was that the culprit could not be a silent mutation. It was argued that it would go against the general observation of most researchers in the area of molecular biology. Reinforcing that erroneous assumption was the discovery that many silent mutations in various species were preserved over a long period of time. In addition, in many species the consequential change made protein more efficiently. However, there was a pronounced and significant exception; it did not apply or did not work in humans. As the human genome sequencing technology advanced, scientists began to question their erroneous assumption. It was an assumption based on scientific observation and logical conclusion. By now, you may be fully aware and most likely convinced how efficient processes of protein synthesis have developed in our body. For instance, you may recall that in RNA chains, the letter U for uracil is substituted, meaning changed for the letter T, thymine, in DNA chains. During this complex process of protein manufacturing there is a letter or nucleotide change for the good or benefit of life in general. Simply reasoning it out, the information encoded in nucleotides is converted into the language that builds amino acids which in turn will finish up as a protein. During the whole process of manufacturing proteins there is a set of rules and steps that should not be broken. This set of rules governing DNA-RNA is widely known as the genetic code. Among the rules-code- is the formation of messenger RNA, and its editing process (splicing non-coding DNA), transfer RNA that bears and carries amino acids to its destination, the ribosome. The job is not as easy as many people think it is. Take, for instance, the amino acid leucine. It has six codons, all coding for the same amino acid. The codons are: UUA, UUG, CUA, CUC, CUG, and CUU. On the other hand, you have the amino acid tryptophan which has only one codon, namely, UGG. You have the amino acid serine with six codons, and tyrosine with two codons. In all, there are 61 codons codifying for the twenty amino acids. In addition, there are three stop codons. As

I said earlier, this set of rules cannot be broken or altered; otherwise we will end up with multiple health problems and diseases. Do not be frightened, we already told you that the cell possess a self-correcting system to ameliorate a problem. Just a few more observations on this subject may be helpful to some students who like to go a little further on gene mutations and related issues in biology. A single letter change to the code we have talked about which ends up as a healthy-efficient protein, known among neuro-scientists as point mutation, can provoke a changed codon forming a wrong amino acid. In professional language this wrong amino acid is known as non-sense mutation. Consequently, the resulting product would be a shortened protein. Therefore, a single letter change in the chains we are writing about can cause several problems. Among those problems we include a stop codon which will provoke a change in amino acid culminating in a longer or different protein. No wonder scientists in this field of molecular biology could not point with certainty the locus of problems and come out with the correct conclusion.

In retrospect, scientists had begun to challenge the prevailing theory on silent mutations about thirty years ago. By the middle of the 1980s, scientists began to realize and accept that silent mutation could intervene and improve protein synthesis, at least in single cell organisms. With new and improved research technology, scientists began to discover that cells had preference for some codons, meaning that cells did not treat codons on equal basis. It seems that cells were employing protein synthesis efficiency selectively. Just as we tend to care for and breed animals and birds that satisfy our curiosity, needs and interest, recent research found out and came to a rational conclusion that mammalian genes tend to favor certain codons over other codons.

To please my tormentor who does not stop asking me to repeat things for him, I will say that the genetic code that we have referred to many times, is nothing more than a set of rules or code in a cell. This set of rules or code dictate or govern how information encoded in genetic material such as DNA or RNA sequences will turn out as amino acids, and ultimately, come out as proteins from a ribosome. The in- between processes of nucleotides, triplets, transcriptions and translations, making a stop at a ribosome, are indispensable rules which cannot be avoided or broken. These are in-between steps the gene code has imposed on the process of building proteins before the final product will be delivered to specific organs within your body. You owe it to yourself to learn how you are constantly re-building yourself to keep your body beautiful, handsome, healthy and intelligent.

A BUG INSIDE MY BRAIN?

Just a few words to get my brain clear of something that has been bugging me for a while now. During this essay, it seems I have picked on the translation process as the bad boy or villain in the gang responsible for many of our diseases and disorders. Please, believe me. I am not prejudiced against translators or anyone else. On the contrary, we recognize how difficult it is to get an abstract concept and decode it into understanding everyday parlance. Sequencing the human genome and learning about all the debris, markers, mutations, modifications, intervention by short segment of genes, etc. makes me admire their job with awesome respect. Discovering and translating for us short nucleotide sequences responsible for flagging or alerting boundaries of an exon up to cellular editing instructional machinery, is awesome and worthy of praise by all of us. This editing process becomes even more interesting while you find SNPs (short nucleotide poly-phormisms) engaged in, or influencing this already complex process of protein manufacture. During this scenario there is no room to blame translators or researchers. It is the gene code itself as expressed in DNA and RNA. We have said several times that brain matter inside your skull is basically composed of glia cells. Some experts in the field claim that over 90 percent of your brain matter is not involved in making good and useful decisions for you. Few scientists who like to speculate on this subject claim that around 93 percent of our bodily flesh, but in particular, our brain, is a left over from evolution. The human brain can be seen as a huge mass of cells, meaning glia cells, occupying much needed space and consuming a high proportion of sugar and oxygen that communicating neurons could use for the benefit of our body. A one 100 billion neurons are busy working for you 24/7 while around 850 billion glia cells seem to be taking a long nap.

However, some glia cells out of that huge number of cells are part of our brain blood barrier. They are performing as inspectors of blood content delivery to all cells in the brain. There are other equally important jobs glia cells are responsible for. However, as far as I am concern, it is the comparable small number of neurons that keep me busy learning about myself. Bugging my brain is consciousness and unconsciousness. In my book, An Episodic Toxic Memory, a survival story, I attempted to explore both subjects, but unfortunately, it became short of a satisfactory answer. We hope neuro-scientists can tackle the issue and provide a satisfactory explanation for most of us. You may suggest that we should begin with single cell organisms, which we have. Peter, sitting at the end of this working table, may rightfully argue that single cell organism do not have the complexity of a human brain. John, a faithful believer in a supreme being, intervenes in this discussion by arguing that science could never decipher the mystery of the brain. The brain is the home for man's soul. It cannot be placed on a petri dish, test tube or seen through a PET or fMRI scanner. This is an argument I prefer not to get involved in. We have attended several meetings on the above subjects. We have agreed to have more meetings, but opinions remained as far apart as we began the first day.

You are unique in our solar system

We have no doubt that we are the most beautiful and intelligent creatures inhabiting our solar system. Cells making your body took many years to develop and build the most complicated and beautiful organism existing in our solar system.

We are an intelligent species with vast knowledge about ourselves and everything existing on planet Earth. We know a lot about our sister planets and sustaining star, our sun. Our astrophysicists have calculated that we have a few more billion years to live before it will change into something else. We have developed sophisticated tools to examine and study ourselves including the most complex organ in our galaxy, our brain. This is a little far fetched extending our imagination, but at present time no one has challenged it. Just a decade ago, we completed the sequencing of the human genome, something that not even gods had attempted to do. At present time, we are engaged in a scientific research race to find deficient genes, with its variants and DNA markers that are linked to diseases. In the past, we pleaded gods, shamans, healers of all kinds and even religious icons for help. Today we have invented first rate laboratories and sophisticated machines to probe our atoms, molecules and cells of our body. We have made progress, but much remains to learn about the intricacy of cell life in our organism. Naive promises were made to the public, especially to individuals and families with chronic degenerative diseases hoping for a miracle pill. They were hoping to end their miserable and painful life by new forms of scientifically proven drugs or therapeutic tools. Locating the gene or genes responsible for the etiology of our diseases has been quite elusive. However, we have made significant progress in all areas of medical technology and science in general if we look back fifty or sixty years ago.

Committed and enthusiastic, but naive

Rather naively, but fully committed and enthusiastic, we were hoping that clear cut diseases like Parkinson, Alzheimer, multiple sclerosis, and to a lesser extent, even schizophrenia, bipolar and cancer, would be eradicated from planet Earth. During the 1990s many predictions were made by lay people as well as some scientists. However, it proved to be poor judgment and unproven science. Hardly any substantive treatment progress has been made or found on the above mentioned human maladies. Further hampering our disease hunt scientific race, funding was cut in many areas of research. Moreover, concerned citizens negatively influenced congressional legislators over stem cells research monies. Besides premature expectations and limited funding, at the time we believed the scientific community working in the sequencing of the human genome did not expect to find that most of our genome is composed of non-coding genes. The problems it presented to them continues to be unsolved. Even after discovering the so call junk DNA, we did not know what to make of it. A few theories have been advanced, but the interrogation remains largely unanswered. Similarly, DNA markers, epigenetics and much more, have slowed down our race in the hunt for the origin of most human diseases. There is a mountain of specialized research which needs to be translated into therapeutic tools and drugs. We applauded Francis Collins from the NIH for taking wise steps in this direction. Mel Graves, a cell biologist writing for Nature said: "We fooled ourselves into thinking the genome was going to be transparent, a blue print, but it is not. Instead, sequencing and other new technology spew forth data, the complexity of biology has seemed to grow by order of magnitude. [61]

In the same article, (04/01/2010, # 7289), Eric C. Hayden wrote, "Much of the non-coding DNA has a regulatory role, small RNA of different varieties seem to control gene expression at the level of both DNA and RNA transcripts in ways that are still only beginning to become clear. (Please, take note of the words seem and beginning to verbalize his opinion). The author of the article, E.C. Hayden, quoted another scientist, a mathematical biologist who said: "Just the sheer existence of these exotic regulators suggest that our understanding about the most basic things- such as how a cell turns on and off- is incredibly naive." In contrast, some of our television cartoons make it easy for us to continue to entertain unrealistic promises. A TV song goes like this: Triplets nucleotides UGG codes for tryptophan; it loves its neighbor AUG methionine with three letters each. The song continues: Not far away we find leucine that can come out in six different triplets. All these letters are translated into amino acids that will become proteins for my body. I found the song very interesting and educational. But, I could not help stop thinking about how far removed from the reality of the process it is. Protein synthesis does not come as easy as presented in the classical DNA-RNA-protein dogma by our double helix discoverers icons: Watson and Crick.

Protein encoding problems

The song goes on naming more nucleotides and amino acids. She deserves our admiration and recognition for introducing the code of life and thus, biology and genetics, at an early phase of life. I wish I had a science minded teacher during my early school

---

[61]-Nature, Vol.464, # 7289, 04/01/2010, p 664.

years. However, the reality as we have found out at present time is not that simple. The genetic instructions of our complex organism have proven to be more difficult to decipher than we had anticipated. Nucleotides encoding for our proteins encountered a myriad of problems still largely unknown. Take, for instance-introns- the non-coding portion of our genome. These introns have to be identified, cut and removed from the mRNA precursor before it can be considered for protein synthesis. Both, the non-coding genes and the splicing processes are not fully understood to be able to relate to disease genesis. Furthermore, it is easy to talk about replication, transcription and translation processes during protein manufacturing. However, connecting it to a syndrome or clear cut disease is basically an unrealized dream in the horizon of medical research. Nonetheless, DNA technology has become a strong and helpful ally of medicine and health in general.

For us, removing introns from exons in pre-mRNA is seen as a wonderful trick of our cells. However, it is in need of further research in the protein manufacturing process for the benefit of disease- genesis clarification. If we consider pre-mRNA's first step and all the steps in-between until we get mRNA and tRNA transported to a ribosome culminating in very specialized proteins in the prefrontal cortex, it is a very long, long trip. It has to be celebrated as a wonderful achievement by our primordial cell. The same thing can be said about specialized proteins in major organs of our body. Some researchers have voiced their concern identifying a loci or a simple step during protein synthesis that would end up in a deficient organ or tissue. This concern stands out without considering many other causal factors in disease genesis. For instance, our laboratory companion, the mouse, is extremely helpful and relatively easily engineered to carry out our experiments. However, a schizophrenic or bipolar mouse, despite its almost similar genome, is not a human being growing up with early symptoms that later on will be developed into a full blown schizophrenic. The intricacy of a gene generating a multitude of messages has us wondering about a simple presentation of our genome and undelivered past promises. We are working very hard, but the task ahead is a challenge in need of bright brains and commitment. We have been focusing our attention on rare variations that applies to population worldwide. How about relatively new mutations, modifications and adjustments maneuvers that boost up human evolution? Those mutations and changes might be in the process of making proper and adequate accommodations in our brain. Our hippocampus and prefrontal cortex are composed of groups of neurons under constant, persistent and damaging stressful conditions. Residents of urban areas are highly exposed to stressors not imagined by our own great grand-parents. Our entire body, but in particular our brain, is making relevant adjustments to multiple demands to be able to survive. How exons and introns interact in critical regions of the brain to come out with appropriate signals for a timely and efficient response is another question that keeps on bugging my brain. Introns and glia cells are demanding more attention from all researchers. In addition, we are not only referring to a specific locus in a cell or organ. We are contemplating the influence played by an incomplete or wrong signals as it relates to healthy and not-so-healthy genes. We have in mind syndromes like schizophrenia, autism and bipolar depression. We see it not as a cut and dry disease, but a mountain of symptoms contributing to a very painful brain disorder.

David Lewis and schizophrenia roots

Leading researcher, David Lewis, has spent the past two decades exploring schizophrenia's developmental roots. He runs one of the most comprehensive and sustained attempts to explore normal and pre-schizophrenic adolescent brains. Dr. Mayberg at Emory University described it as, "One of the smartest, most creative and most promising angles I know of on schizophrenia." Dr. Lewis focused his attention on the prefrontal cortex we mentioned several times during this research essay. On above paragraphs and chapters, we have shown you how this region of the brain is connected. It receives a myriad of projections from the limbic system including the amygdala, hippocampus as well as sub-cortical and neo-cortical input. It is often called the CEO of the brain, but it seems partially true because it is dependent on emotions coming from the limbic system to make wise decisions. Furthermore, you have the life saver amygdala that has a mind of its own when it comes to protect you from immediate danger. Dr.

Lewis' work examines the relationship between two types of cells: pyramidal neurons and chandelier cells. He thinks that chandelier cells fail to cultivate pyramidal cells during childhood or early adolescent-hood. The communication needed in this region to develop a robust connection, leaves the prefrontal cortex incapable of coordinated firing and synchronicity. Tom Insel from the NIH says Lewis' model provided something this field really needed: a framework for linking observation at the molecular, cellular and system levels. [62] Coordination and cooperation among researchers, is paramount to be able to tackle syndrome like diseases. With brain diseases and disorders, we hope researchers do not fail to include the functions of the majority of its cells, glia. Astrocytes seem to be eavesdropping on neurons at all time. Oligondendrocytes are responsible for myelin on neuron's axons, among other roles. There are many brain diseases connected to myelin deficiency.

ATOMS AND MOLECULES IN ACTION

When we observe the multiple organelles within a cell, we wonder how atoms and molecules interact with each other to function normally. Molecules are constantly bouncing against each other in an aqueous environment in animal cells. The same must hold true for atoms, but this elusive and microscopic particle of every element is very hard for us to advance an intelligent opinion. Atoms and sub-atomic particles are constantly traveling at huge speeds through, above and under me. You are not exempt or free from this constant bombardment of atoms and subatomic particles either. Let physicists do the explaining for us. However, collision among molecules and atoms of all sorts within and outside our cells produce force, meaning energy. My friend Pete was wondering if the abundance of force floating around molecules of our basic elements like hydrogen, carbon, oxygen and nitrogen are not overly exposed to degeneration. When they are degraded into their basic atomic components they may provoke some of our diseases. This intelligent observation is an interesting one and deserves serious consideration by concern researchers. Of course,

---

[62] -Nature, Vol. 468, 11/11/2010, p. 155-56.

there are multiple chemical reactions going on inside our cells necessary for normal function. Our body is in a continuous process of self-renewal (apoptosis and cell division and multiplication), but we are referring beyond and above our normal birth and the aging phase of our life. Or, is it naive for us to separate one from another? We are thinking in line with our basic molecules during the forming of covalent bonds. These bonds are many times stronger than floating molecules or atoms. Our macro-molecule DNA with the assistance of another equally powerful molecule, RNA, is of paramount importance to you and I. Both molecules are engaged in the production of proteins called enzymes. These enzymes are responsible for the making and breaking down covalent bonds. In other words, these enzymes can tear down the union of indispensable units of our organism that are built to tolerate a lot of tension and stress under critical conditions. They can come under attack by a protein-enzyme coming from an incomplete or dysfunctional gene. This is a very valuable truth to remember while keeping an eye on our health. There are many genes in our genome called introns that at present time, we do not know their function or functions. There are more non- coding genes than protein producing genes. You already know that introns are cleaned out before messenger RNA is ready to be shipped out to become part of a protein. However, you also know the multiple problems and failures a protein encounters before its final step is completed.

Covalent bonds

If you allow us to go back for a little while to try to make our point clear to you, we will appreciate it. Let's begin by saying that covalent bonds are made when atoms share electrons. You may recall your elementary school teacher telling you that atoms are composed of protons with a positive charge and neutrons which are electrically neutral. Protons and neutrons are located in the nucleus of an atom. The teacher also taught you and I that atoms have electrons traveling around the nucleus in an orbital fashion. Electrons carry a negative electrical charge. Generally, atoms have two types of sub-atomic particles in the nucleus, namely, protons and neutrons. Protons are electrically positive charged. Electrons carrying an electrical negative charge travel around the nucleus of the atom in an orbital fashion. Neutrons are electrically neutral. Hydrogen has a positive charge proton with an electron, but no neutron. The most abundant atom in our body is the hydrogen atom. It is the lightest element that became part of our organism. Hydrogen atoms mix easily with other atoms. A basic law of physics to remember is that the electrical charge possessed by each proton is exactly equal, but opposite by that carried or possessed by a single electron. This positive/ negative charge of protons and electrons would make the atom electrically neutral. The mass of each neutron in the nucleus is equal to each proton. It is another basic rule you must remember.

Another abundant atom in our body is carbon atoms. In the nucleus, a carbon atom carries six protons which are electrically positive charged. In addition, there are six neutrons   carrying an electrical neutral charge. Most of carbon atoms on planet Earth are composed of six protons and six neutrons known as carbon 12. Neutrons contribute to the stability of the atom, but the sub-atomic particle that definitely determines the element each atom belongs to, is the proton. There are carbon 13 and carbon 14 known as isotopes. Carbon 13 has the normal 6 protons, but 7 neutrons. Carbon 14, a very radio-active isotope, has the normal six protons, but 8 neutrons. The six protons in carbon 12 in its normal condition remain unaltered. If for any reason, atom carbon 12 would degenerate to 5 protons or acquired an extra proton, making it 7, that atom is no longer a carbon atom. However, a couple of pages back we talked about covalent bonds. These bonds are very strong and do not break down that easily. There is a significant difference between sharing an electron between two atoms that fulfill mutual specific needs for each other, and releasing or donating electrons to atoms or molecules. Besides strong covalent bonds, there are hydrogen and ionic bonds that we will not address here. While Watson and Crick were working on the structure of the double helix, one of the mind bugging issues was how the bases inside both backbones came together. Nucleotides from both backbones were united by hydrogen bonds, a weak bond. This hydrogen bond would facilitate the opening of the double helix for DNA replication and transcription. We brought this up in an effort to make you aware of the many complex processes our proteins, organs and tissues go through before they become functionally normal in our body as well as yours.

Nowadays, some individuals refer to us as made of carbon chips. What it means is that if we were to do the impossible, meaning, take all the water out of our body, approximately, all the molecules in every cell in my body and your body are based on the carbon element. Carbon atoms are extremely versatile. They can easily form covalent bonds, join other carbon atoms forming large chains and rings for complex molecules. When you look at the double helix, you see the inside bases, adenine, thymine, guanine and cytosine in the form of rings. Carbon atoms are involved in DNA methylation and histones modifications, among many other interventions in protein synthesis and cell normal functions. In the same vein, amino acids are a particularly characterized class of molecules that are distinguished from any other molecule in our organism. They all have an amino acid group (NH_2) and a carboxyl acid group linked to a single carbon commonly called the a-carbon. There is a side chain or molecule known as R which is attached to this a-carbon. This side chain provides the chemical variety to amino acids. We do not shy away from repeating that proteins are manufactured from 20 amino acids that are vital to life on Earth. It is not intended to be a joke, but amino acids and proteins are parts of your body in sickness and in health.

Just as amino acids are the subunits of proteins, nucleotides are subunits of DNA and RNA molecules. Basically, a nucleotide is a molecule compound of a nitrogen- containing ring compound linked to a five carbon sugar. The sugar in this nucleotide can be either ribose or deoxyribose; it carries one or more phosphate group. You see it clearly in the double helix as per Watson and Crick's classical illustration. Although the word nucleotide is commonly used to define or refer to the famous letters ATCG, it also refers to the phosphate-sugar backbone segment linked with each base pair on both sides of the double helix. Nucleotides have additional functions we are not equipped to deal with here. In summary, you have the ribose containing molecule RNA and the deoxyribose molecule. The RNA molecule spews out uracil for an exchange of thymine and DNA equally spews out ATCG.

Amino acids are a very particularly characterized class of molecules that distinguishes itself from any other in our organism. They all have an amino group, (NH_2) and a carboxylic acid group linked to a single carbon atom, commonly called the a-carbon. There is a peptide chain, known as the R chain, attached to this a-carbon atom, thus providing the chemical variety to amino-acids. These amino acids molecules will be involved in the manufacture or synthesis of proteins. And… these proteins are part of your body in sickness or in health. Just as amino acids are the subunits of proteins, nucleotides are the subunits of our two great molecules, DNA and RNA. Basically, a nucleotide is a molecule composed of a nitrogen- containing ring compound linked to a five carbon sugar. The sugar in this nucleotide can be either ribose or deoxyribose; it carries one or more phosphate group. You can appreciate it clearly in the backbone of the double helix with the bases in between. Repeating a lesson is a carry over from my teaching experience.

Although the word nucleotide is commonly used to refer to the bases, it also refers to the phosphate, sugar backbone segments linked with each base pair on both sides of the double helix. Nucleotides have other functions we are not addressing here.

Teachers love to go over and over again their lesson so it becomes long-term memory on your brain. Do not blame us for being repetitive. Consider it as an innocent attempt on our part for you to remember us. My teacher used to hand out hand written declarative sentences like: On a weight basis, macromolecules DNA and RNA are by far the most abundant of the carbon-containing molecules in a living cell. To make her point even more clearly to us, she used to add that all cells in her body were made of dear carbon atoms. Carbon molecules are the most distinctive properties of all living things. At the time, I was wondering about my soul, oxygen, nitrogen, hydrogen and other elements that exist in my body. However, she did not emphasize it as much as she did carbon atoms. I remember her as a caring and dedicated teacher who accomplished her goal of creating long term memory in my brain. She hammered on DNA and RNA molecules on us because she knew, among other things, that both molecules are engaged in the formation and storage of memories. It is in addition and above transmission of hereditary information. And, involved during this storage and transmission of information, are our inherited disease propensity risk we have been writing about. It is truly a long journey from our ancestors thousands of years ago to the present. We have survived not only the jungle, but the armies of microbes, plagues and natural disasters of all kinds. In great urban areas, when we ride public buses and subways, we are breathing the same air that is infected by sick people riding with us. A secondary problem coming out of sharing the same air trapped in subways and elevators is that not everybody has developed the same defenses against bodily intruders like viruses and bacteria that will make us sick. Let us say you were born and raised out in the country, not far from the ocean like me. We were used to eating fresh fruit and vegetables picked up from our small farm. We ate a multiple number of beans and roots accompanied by home grown chickens, goats and cow meat. The meat was usually dried out and salted at home. My mother, like her father, was a locally well-known botanist who kept a medicinal garden at home for family and neighbor's needs. Although tobacco was one of the main sources of income in our home, none of us learned to smoke cigarettes; it was exported up North. I do not remember going to a medical doctor, except a dentist, until I was nineteen years old. I had left my country home and move to a northern urban area riding trolleys and subways. I was exposed to alien micro-organisms my body had not built adequate defenses for it.

Notes from my teachers

From my teacher. I carry notes reminding me that the DNA polymerase is trusted with the responsibility of manufacturing new DNA strands by using the "old" strand as a template. This self-replication process provides the mechanism for the addition of new double strand of DNA molecules. DNA replication produces two double helices from an original DNA molecule.

Each of the two strands of DNA is used as a template- a mold- for building complementary strands. On top of a blackboard,

132

she had written: Most genes are short segments of DNA encoding a single protein. The DNA in a gene is not used in its totality encoding a single protein. Short stretches of DNA become busy identifying, determining timing, place and size, among other jobs, in protein manufacture. We do not tire repeating that you have a very wise DNA molecule. Your genome is a large and effective information storage vault that deserves much respect and awesome admiration. Another awesome attribute of our DNA double strand and self-replication is that when it separates for replication, the information is passed on to the complementary strand.

DNA replication process

The DNA replication process is initiated by a protein-enzyme- that attaches itself to the DNA and opens up,-unzips- and separates apart each strand. It does this job by forcing loose or unbinding the hydrogen bonds between the bases. There are particular gene segments responsible for signaling where the replication process begins. During the whole process of replication there is a very wise enzyme known to us as DNA polymerase that is busy directing this complex operation. It synthesizes the newly formed DNA strands using one of the old strands as a template. This wise DNA polymerase commands the addition of nucleotides to the growing DNA strand. The term old does not refer to time or age. It refers to one of the original strands that will serve as a mold for a complementary new strand. For this very reason, each strand of DNA molecule becomes a template for a complementary strand. This enables the cell to copy its genes before passing them on to coming generations. This wonderful process which is carried out within the cell must be done extremely fast and accurate. The beauty of this replication process is that it may include many thousands of nucleotide pairs. This replication machine which is basically composed of proteins, manufactures two complete double helices that are identical to the original DNA. This replication process involving thousands upon thousands of nucleotides must do it within a minimum of errors. Take a minute to think about people who live to be a hundred years old. Those molecules responsible for replication, transcription and translation are active without taking a week or two vacations a year, deserve to be respected and praised for a job well done.

Your chemistry, DNA plus

Getting to know your chemistry, meaning your DNA, you will learn the rules that govern all forms of life on planet Earth. It is the chemical soup that gave rise to an RNA and DNA world, that in relative harmony with another wise molecule, the mitochondria, has produced the most intelligent organism in our solar system. Our laws of physics, chemistry and biology seem to work in unison with, and applicable to the rest of the universe. No matter which element of our body you take, it is composed of atoms. And... atoms of various elements are flying or traveling in space in the form of clouds-gases-and become attracted to each other forming celestial bodies like stars. And stars, like humans, are in a constant dynamic process of birth and death ever since a big bang took place somewhere around 18 billion years ago. I am not good at figuring out time, but it happened many, many years ago. Some stars go through the process of implosion-explosion becoming what is known as a supernova. It is a type of recycling process in which the old star goes through a cycle of self-consumption to appear anew in atoms and molecules in another space and time. A few billion years ago, our star, the Sun was born, and in another few billion years more, you can rest assured that it will die. You and I will not be here to witness it. However, the same atoms, the same energy that exists in both of us at the present time "alive", will forever exist; except if it falls prey to a black hole, according to most modern astrophysicists.

Tiny particles inside the atom

In an atom of an element there are tiny particles and waves responsible for chemical and atomic processes we are beginning to understand and use for our own benefit. Likewise, in our bodily cells we have chromosomes that are structures within the cell nucleus where genes are carried or stored. There are more genes than chromosomes. It is not only a scientific truth, but a logical conclusion. Consider the genes you must carry for the color of your eyes, skin, hair, and even your height, to some extent, is determined by your genes. These genes you get from your parents are filed in your DNA, great and wise molecule, waiting expression during mating. The genes you have inherited from your ancestors do not fade away, fuse or blend in a fertilized egg. They remain unaltered, or as Gregor Mendel called it, particulate, meaning that those very particular genes are not lost after fertilization. This wise and humble dedicated monk, G. Mendel, discovered the unblending quality of genes around 150 years ago during his spare time while teaching at a monastery. We can say without an ounce of doubt that he did not have a first rate computer to work out his mathematics and necessary computations for his experiments. Monasteries during the second half of the XIX century in Eastern Europe, Czech Republic, were not known for the abundance of anything. However, writing paper and pencil or pen, he must have managed to keep in his dormitory room. The monk and geneticist Mendel, knew that his discoveries (genes) were very important, but his fellowmen did not appreciate it. In the year 1865 he gave a lecture on his work at a local medical society followed publication of his discoveries in the society's journal. Unbelievable or not, his findings remained unknown to the scientific world and public in general until the first decade of the XX century. Mendel had proven scientifically what farmers had been practicing for thousands of years. Farmers had been taking selectively the best and desirable offspring (gene traits) of animals, birds and plants to improve their stock. It appears to me that Mendel's work is just recently appreciated during and after the sequencing of the human genome. We are learning to engineer genes for medical and food research for the benefit of humanity in general. A great section of our essay has been centered on DNA, genes and proteins. Gregor Mendel was born in 1822. He did not live to see the fruit of his work appreciated beyond the walls of his monastery. For me and many others who love his work, Mendel is another hero who deserves to be paraded along the Heroes Canyon on Broadway in Manhattan. Without insulting anyone, the science of genetics was born by counting round and wrinkled peas by a monk who was not known by his intellectual acumen.

My inquisitor is asking me to include G. Galileo, Marie Curie and Rosalind Franklin on the list of outstanding scientists to be selected for the Heroes Canyons. You should have a voice in the selection of our heroes. We did not mention Francois Jacob and Jacques Monod because they were paraded along the Arc de Triomphe in Paris.

Gregory Mendel based his conclusion by observing hereditary determinants-genes- as phenotype, meaning their outside appearance or expression. With pea plants he observed and counted spherical and wrinkled peas. He began his work by selecting and observing pure spherical and pure wrinkled peas. He paired or fertilized a few of these so called pure breeds. The first generation of fertilized peas, which are clasified by geneticists as F1, came out all spherical. This is phenotype or external appearance of each pea he collected. Moreover, the scientist monk mated or fertilized the offspring of the first and subsequent generations of his selected peas. He found out that they came out spherical and wrinkled peas in a ratio of 3 to 1. The spherical determined traits in peas he called dominant. The wrinkled peas, he named it the recessive hereditary determinants. What Mendel called hereditary determinants is known to us as genes.

Without tools to look inside each pea to examine and analyze how hereditary determinants work and remain "particulate," Mendel had half opened the door into reading the book of life. Future scientists became interested in discovering not only how and where are hereditary determinants-genes-formed and stored. Scientists wanted to know the alphabet or letters which would enable them to read how peas, flowers, plants, animals, and above all, how human beings are being built from apparently, two simple molecules. We do not tire of thanking a simple and humble monk for showing us, among other things, the outside expression of genes. At present time, we have become engaged in an investigation race or hunt for all sorts of genes. These include mutated genes, disease provoking genes, rare gene variants, comparative gene evolution among species, and many others. We wanted to identify, characterize and look inside genes and use its million years old code for our benefit. After a short hesitation whether hereditary information was stored in proteins, amino acids, or in similar chemical structure within the cell, scientific based experiments proved without any doubt that the genetic code resides in the DNA molecule.

Among the big players during this hunt for genes was L. Pauling, known to many scientists as the world leading chemist during his lifetime. Besides Pauling, there was Rosalind Franklin, Maurice Wilkins, and of course, our intellectual icons: J.D. Watson and Francis Crick. We are not overlooking the contribution made by the world best known neurologist, Santiago R. Cajal and many others who would be impossible for us to mention here. Cajal made it clear for us that our brain is not a long sausage coiled around itself. He set it clear that our brain is composed of neurons and glia cells that are independent of each other. Neurons communicate with each other through a space between them known as a synapse. They send and receive messages among themselves using chemicals called, neurotransmitters. For further information on the subject see the following books by this author: Parkinson, Alzheimer and schizophrenia, as well as, The Brain, the world's most wonderful organ. Returning to Watson and Crick's double helix and, " it has not escaped our notice that the specific pairing we have postulated...", pulled out biology from cataloging and classifying animals, plants and birds into a scientific race just rivaled by that of the atom. From that date on, Presidents and Prime Ministers on both sides of the Atlantic appeared on the front pages of newspapers and scientific journals with Nobel Laureates and pioneers in biology.

Watson and Crick had set in motion a race among scientists around the world. They wanted to know how amino acids were manufactured and built into proteins. The genetic information Mendel was referring to was found to be encoded in the sequence of nucleotides-the letters of DNA. However, a clear and accurate look at DNA and how it determines its nucleotides formation still needed some extra work and delicate investigation. At the core of this scientific investigation and intellectual curiosity lies how genes work and behave at molecular level. Genes, proteins, diseases, drugs to cure it, and research tools seemed to have occupied the mind of scientists around the world. Out on the horizon, far away and hardly conceptualized by most people, there were scientists attempting to understand gene behavior and the sequencing of the human genome. Questions such as: How do we look at uncoiled or unwraped DNA strands and the chromosome itself, were academic debates with very few scientists engaged in laboratory work. We wanted to know how long strands of nucleotides, coiled up in beads like structures with multiple amino acids, could conserve unblended, Mendel's particulate quality.

DNA TECHNOLOGY AND RESTRICTION ENZYMES

This race in biology was turned into a DNA-RNA technology revolution. Biologists, chemists, physicists, mathematicians and computer engineers began to see the advantage of working together. Among the first DNA revolution technology tools we came across were restriction enzymes. It meant scientists could cut DNA molecules segments using the newly discovered restriction enzymes. DNA segments cut using restriction enzymes, meaning a gene, or clusters of genes scientists were interested in, were kept away, isolated for study, analysis and purposeful scientific use. There was a scientific goal in mind to achieve using this new technology. The next step scientists took upon themselves following restriction enzymes discovery, was making use of a wise protein, an enzyme dubbed ligase. This ligase enzyme could be used to paste or glue together DNA sequences- genes-for whatever reason the researcher had in mind. According to J.D. Watson, there was a young scientist gifted with a business mind besides his enthusiasm for the genome sequence. His name, "Herb Boyer, was already an expert on restriction enzymes in an era when hardly anyone had heard of them," wrote Watson. Well, at this early stage of DNA technology we had mastered two enzymes to do work for us. There were two enzymes we could use to cut and splice genes. Was it not a wonderful discovery! Mendel played with phenotype expression of genes, but we were touching and manipulating the genes themselves! Furthermore, the technology of cutting and gluing genes together became known as a plasmid. A plasmid is roughly like a CD for the recording and storage of information. By luck or coincidence, there was another brilliant scientist, Stanley Cohen, attending a conference on plasmids in Honolulu in 1972. So, by the grace of curiosity and scientific interest, "the first among plasmids," S.

Cohen and Herb Boyer, the restriction enzymes star, realized that together, they had a DNA technology nobody else knew about.

Besides their interest in genes, their chemistry seems to have worked very well, and together, they established the first recombinant DNA company in the world. This was the world first genetic engineering establishment. It was done on the west coast of U.S.A. This game that consisted of playing around with genes and enzymes was done with small animals. The big job, the sequencing of the human genome was resting, if at all, on the drawing board of optimist biologists and scientists.

Boyer and Cohen's discoveries, and DNA technology tools were not a small thing to go unnoticed by the scientific community around the world. They could play-engineer- with bacteria and small animal cells, make cuts with restriction enzymes and put it back together with a ligaze enzyme. Cutting and joining together compatible segments of genes gave birth to another novel idea, recombinant DNA technology. One dear parasite in our intestines, E Coli, proved to be an excellent subject to work with. It is cheap and it hardly demands much of you. How to transfer DNA material from one organism to another was the next challenge. Well, we can make use of plasmids to do the job for us. The question that lies ahead is how would the host receive its guest? When big foreign dignitaries come to Washington we display a big red carpet for them. Not all of them deserve such a privileged treatment, but politics has its own rules of behavior. You already know how one organism tends to defend itself and reject what it considers invaders or intruders. We are referring here to our defense system or immune system. There is a group of defensive cells which has its origin in the bone marrow, with T cells maturing in the thymus. It is a survival system which has defended you and I for thousands of years. It mobilizes thousands upon thousands of T and B cells to keep you alive when a toxic invader tries to break into your body. What I am saying is that the solution to recombinant DNA technology rejection problem already exists in your organism. Your brain has its own brain blood barrier to filter out undesirable organisms. Boyer and Cohen had the solution at hand using a plasmid, meaning test tube experiments. Cuts on sequence segments made by plasmids would be glued together with a ligaze enzyme. At first glance, it did not provide much help to Boyer and his plasmid pioneer. The ligaze enzyme was doing what was already known to the researchers, gluing stretches of plasmid together. However, the opportunity to come out of the laboratory with his test tube in hand crying out, eureka, eureka, meaning, I found it, I found it, was not far away. During a following test, it was observed that the gluing enzyme would manipulate a cut plasmid to incorporate segments of DNA from neighboring plasmid, thus creating a hybrid. It was in Cohen's hands to use his already proven technology to transplant the whole plasmid inside a bacterium for multiplication purpose. Resistance to the newly added guest was overcome using already available antigen resistance. The next step would be the creation of a hybrid from two different species. We have an abundance of flies and birds we can capture and use in laboratories for the benefit of mankind. I remember crows coming to my father's corn field to eat corn on the cob despite our strong objections to it. My father spent money and time planting and cultivating corn while crows had a feast at our expenses. We could have captured a few crows, extract a few cells, cut DNA sequences using restriction enzymes, glue it to a bacteria plasmid using a DNA ligaze enzyme, and bingo; we came out with a recombinant DNA molecule. It looks very simple the way we have described it in here. Perhaps after we practice it five or six times, it would be a simple job for all of us. It would be just like eating a piece of cake.

TV shows and DNA finger printing

Up to this point, most of the work we have been writing about had been done with bacteria. The next step would be with higher animals having a similar genome to ours like our little brother, the mouse. Recombinant DNA technology is and will be an excellent tool for the discovery of DNA mutations that are responsible for inherited diseases. It is already a first rate tool in medical research and therapeutic tools beginning with insulin for diabetics (Boyer and Cohen) as well as blood clotting proteins. Another tool we cannot ignore in this essay is DNA finger printing, although the finger is hardly involved in the game. Sometimes I enjoy television shows presenting two or more young fellows claiming or disclaiming paternity privileges or monetary contribution to the mother of a child. At times, I believe the TV show host presenting two or more males with the same claim, is just a charade or stunt to arouse the spectator's feelings and improve its show rating. However, DNA technology has come out to be the jury and judge in such cases declaring who is the father of the child in question. DNA identity is 99.99 percent accurate despite the mother allegations that she slept only two times with the other young fellow. Besides paternity issue, I like to watch crime stories when DNA technology is involved in solving a crime.

Attempts to hide DNA finger print.

There have been criminal cases extremely difficult for police officers and detectives to solve and identify the criminal agent. The delinquent subject has been a cold and calculated person who has carefully considered all possible clues he might have left behind on the scene of the crime. I hear police officers and detective call him a professional criminal whose specialization is eliminating all traces of possible identification.

Among the things he might have done are covering foot tracks in multiple ways besides covering his face by wearing a mustache. Some times the criminal agent paints or enlarges his eyebrows, modifies the shape of his nose or alters his voice among many tricks he has learned to evade capture. He may appear to be the perfect serial killer, rapist or bank robber. He is placed on the list of most wanted criminals, and has hundreds of agents watching for a man with specific detailed description of him. He might have changed his phenotype or outside appearance, but his DNA profile is forever, no changes are allowed. This criminal might have forgotten that his DNA left on a cup of coffee, soda or cigarette butt is enough to put him behind bars after he has seen the awaiting judge.

Leaving behind the serial rapist, bank robber and paternity claimant, we return to our wonder working enzymes. Restriction enzymes cut DNA molecules at specific sites. They know exactly where to go assisted by small segments of genes. Those cells of yours are not stupid at all. They have developed through the years many tricks for self-protection. We are just now learning about them. While our scientists were doing their job investigating the inside of a bacterium, they observed that the host bacterium cut plasmid DNA segments at particular places. The mysterious enzyme (restriction enzymes) cut DNA only at certain nucleotide sequences. This was another bingo celebration worldwide.

They were very particular choosing the sites to do the cutting. Moreover, there was another very interesting factor or piece of the game which had to be looked upon and included in the game. Not all bacteria possess the same restriction enzyme; therefore, some restriction enzymes would cut at different and selected sequence of nucleotide. This discovery was extremely important for the near future. It played a critical role during the sequencing of a large genome. One of the rules of the game practiced by bacteria cutting enzymes is that they choose to cut short, from four to eight nucleotides pairs. The following insight into cutting enzymes was that knowing that there are different bacteria, each species or subspecies would have its own enzyme cutting at specific sequences. These discoveries on bacteria enzyme cutting games proved very useful for future sequencing of large genomes. Restriction enzymes cut at certain sites containing different amount of base pairs. In a properly and adequately prepared aqueous medium or gel, the longest and heavier cuts of base pairs would assemble at different places from the shorter segments of base pairs. It sounds logical and even practical to me that scientists came to that conclusion at the time. Brain power plus dedication and perseverance make them a breed apart becoming Nobel Prize winners. The abundance of bacteria in our environment provided us the opportunity to have multiple restriction enzymes cutting at different sites available to us and ready for use in laboratories. Engineers in this field of molecular biology take, among multiple restriction nuclease available, a specifically tagged one, meaning a given restriction nuclease, from the whole bunch available and ready for a job.

Electrophore Technology

These restriction enzymes will cut at a given DNA molecule at the specified site intended by the technologist. DNA fragment separation from each other is accomplished, in most cases, using gel electrophore. It will separate fragments or base pairs within larger fragments. One relatively easy and new way of doing it, is done by staining DNA with a dye that fluoresces under ultraviolet light. (In the ocean beach in Lajas, Puerto Rico, when the Sun gives way to the night and the stars adorn the sky, you can see thousands upon thousands of tiny fish carrying a fluorescent protein separating them from any other fish. Take a short vacation and visit it; you may be motivated to do some type of research with those beautiful and interesting fish.) If a tiny fish could develop that kind of technology, your brain can do even better. Another model used by scientists is to incorporate radio-isotopes into DNA molecules. Cold Spring Harbor Laboratory on Long Island in New York is a pioneer center of research in this field. Our dear friend J.D. Watson made his first presentation of the double helix there.

Recombinant DNA

The DNA technology is using the most advanced equipment available to commercialize its products in many journals. With tools available to DNA technologists, long DNA molecules can be cut to the desired sizes fragments for an intended job. You just need to join it together with the protein-enzyme we have already mentioned to you. The dogma developed by Watson and Crick that strands will attract each other to form the double helix based on the chemical attraction, you do not need to worry about it. Nucleotide adenine will attract thymine and cytozine will attract guanine. You do not need to worry about this while playing with DNA fragments cutting and subsequent union. This DNA chemical ruling governing all organisms permits DNA from practically any animal source to be joined together. This chemical process allows technologists to have a free hand cutting and putting together DNA fragments from different organisms. The enzyme ligaze can join together two fragments without asking questions about their origin or place of birth. And, the energy to do the job is available and plentiful using another protein, the ATP

143

molecule. It will bring together the sugar-phosphate backbone. Backbone fragments are identified with digits 3 and 5 in most illustrations. The three end of a backbone has the shape of a half moon or the letter C while the 5 end is the full moon or a spherical tennis ball. When joined together they will fit perfectly into each other. You do not need to use a pair of pliers to do it; enzymes will do it for you. During this apparently simple process you will be playing with chemical attraction of base pairs. The energy providing ATP molecule, restriction enzymes and DNA ligaze enzymes will also participate in this game. When these processes are taking place inside your body, you will have supervisors and guardian proteins watching over the job in case there is an error. Your cells are always on super-alert status to intervene when it is necessary to do it. All theses mechanisms fall within the code of life your cells have kept for your own safety. Please, do not destroy it; it took your primordial cell a long time to build it.

Laboratory workhorse, a bacterium

Base pairs in DNA segments are the same in all living organism. Once you join it together as in plasmid, it will continue to replicate as if it were coming from an original and same mother. Basically, bacteria seem to be our best workhorses for duplicating our DNA fragments. Cloning is a process of making identical copies of a subject you want to duplicate. Our laboratory working horse, a bacterium, is an excellent tool to have DNA fragments copied several times fast and economically prudent. Every time a bacterium replicates or duplicates itself will also replicate the DNA fragment you inserted into it.

You may want to or need to replicate human DNA fragments making use of another organism for future use. You may need to cultivate DNA fragments from a person to repair an ear or nose. We would like to share with you that this is not science fiction. When working with DNA fragments introduced in a bacterium, it needs some protection from inside hostile fragments, otherwise, it will become food for the bacterium. A DNA enthusiast at the time (Stanley Cohen) was nearby with a plasmid to be used as a carrier, also known as vector. This carrier model-a plasmid-possess a replication original which will duplicate independently of the bacterium chromosome. During the manufacture of a final product, restriction enzymes and DNA ligaze play a major role on the lasso shape or circular plasmid in recombinant DNA. Recombinant DNA is copied as bacteria divide many times in our scientist's laboratories.

A good trick in this very interesting game is to replicate one's own DNA fragment free of a bacterium harmful content and keep it frozen for future use. The final step during recombinant DNA plasmid is purification of the DNA from the content of the rest of the cell. This process includes the removal of the bacterium chromosome. The final purified plasmid DNA may contain thousands upon thousands copies of the original DNA fragment. There are sophisticated tools to get the DNA fragment clean and stainless out of the plasmid. The thing that impresses me most is that all this new technology was and is stored in your brain cells.

Just one more time

One more time we will continue to elaborate on DNA technology available for research at local individual laboratories. We had stopped at restriction enzymes, plasmid and recombinant DNA. We centered our attention on our icon pioneers: Boyer and Cohen for many obvious reasons unable to mention here. There are other equally outstanding pioneers in the field we did not elaborate on in order to concentrate on DNA and its multiple tools. However, we feel obliged to write something on another Grand Canyon hero, Warner Arber, a Swiss biochemist whose main concern at the time was the following: Why some viral DNA were degraded after it was introduced into a bacterium? Those were days of grand pronouncements on the double helix, amino acids, triplets and protein synthesis. They were subjects of investigation conducive to Nobel Prize and big celebrations in luxurious hotels. Arber's brain was busy at home wondering why some cells were prejudiced against some viruse's DNA. Host cells would selectively destroy some viral DNA. What characterized or distinguished that viral DNA from the rest of DNA? A DNA molecule is the same whether we are talking about micro-organisms like a virus, a bacterium, an elephant or a sequoia tree in

145

California. Arber's sleep was interrupted wondering whether an unknown molecular, chemical process or bugging genie was inside the cell keeping its own DNA untouched but fiercely attacked viral DNA. He faced the challenge with great enthusiasm and an inquisitive mind. He discovered that the genie in question was no other than the now very famous restriction enzyme. His discovery became a first rate research tool in many DNA laboratories. It became the darling enzyme in recombinant DNA.

Restriction enzymes in bacteria cells restrict viral growth by cutting foreign DNA. This is the insightful moment that Werner Arber was waiting for, even when asleep. A lot of science work is done in your head proposing ideas and solutions to problems. Many end up in a laboratory, or rehearsed with peers and co-workers. Further investigation with restriction enzymes showed the DNA cutting is not a random and wild job. It is a sequence specific reaction. It does not cut capriciously and biased. A given or particular enzyme will cut DNA only when it recognizes a particular sequence.

You and I have to agree that it was not a small and insignificant discovery that could have remained hidden among multiple notes and rusty tools. This type of discovery is for you to stop biting on your sandwich, grab your cell phone and contact you friend immediately. You can also attend conferences by these researchers.

My noxious friend and tenant, E. Coli again

We already covered for you Eco R1 as one of the first restriction enzymes to become a star among enzymes. It recognizes the sequence of the following six letters: GAATTC and cuts it from the rest of the letters. Up to now it sounds and looks easy and beautiful. Eco R1 saw the six letters sequence that began with a G followed by two AA and two TT ending with a C and made its cut. Stop and think, do not celebrate anything yet. How come Eco R1 does not cut the genome every time there appears the same sequence of letters? There would be a lot of pieces of genes and fragments of genes from the genome lying around strewn on the

146

table, test tube, petri- dish and perhaps on the coffee pot. W. Arber was doing his science 24/7 and discovered that the bacterium that he was working with, manufactured another enzyme. This newly discovered enzyme did not allow automatic cutting of those six letters we named above. Eco R1 cutting privileges were restricted by another enzyme from the bacterium. This newly found enzyme modified the structure and chemistry of those six letters sequences of its own DNA. This modification of structure and chemistry was done by adding a methyl group to the bases. The methyl group consists of one carbon atom and three hydrogen atoms. The cell itself had invented the cutting restriction enzyme, the sequence cutting interval and the chemical modification to protect its own DNA being cut automatically. When Eco R1 was running around cutting prearranged sequences of letters; it did not recognize the six letters. Otherwise this enzyme would make its cut without hesitation. It looks cool and easy, and should be. Bacteria have all these tools. We, as superior organisms have it and W. Arber proved it for us.

DNA sequence cutting was moving along quite well with new discoveries made on both sides of the Atlantic. DNA technology jargon was frequently heard in bars and restaurants and even in barber shops. For the next step in our DNA technology festival we referred you to, there was a problem that scientists and medical doctors were aware of. It was bacteria resistance to antibiotics. Some people were responding negatively to antibiotics because the medication prescribed did not kill the bug inside the body of the sick person. The bacterium went through a mutation in its genome. The newly discovered antibiotic resistance was provoked by incorporating an alien or foreign small fragment of DNA later called a plasmid. A plasmid has the shape of a lasso or loop. Plasmids have made bacteria their main permanent place of residence. This lasso- loop- structure that lives within a bacterium is not a passive, inactive and uninvolved piece of DNA. It possesses the ability to replicate and pass on to the next step, along with the rest of bacterium genome, while the cell goes through its stage of division. The cell cannot leave it behind or get rid of it during cell division. We have repeated several times, most cell have survived millions upon millions of years. First we learned that bacterium, a cell, survived threats by mutation and secondly, cells have another survival tool, the plasmid. Surprises are not absent during investigating micro-organisms like bacteria. Plasmids were found to pass from a single bacterium to another which originally did not possess it. It was a blessing for the receiving bacterium because it received DNA genes providing antibiotic resistance.

It remained for Stanley Cohen at Stanford University to further elucidate for us the new antibiotic tool, the plasmid. As the story goes, Cohen and Boyer met at a conference in Honolulu in Hawaii. Both pioneers in restriction enzymes and plasmid gave birth to Recombinant DNA. One of the first products of this union was insulin. We enjoy sharing with you these important episodes of great scientific discoveries of our time. This stage of genetic engineering presented serious questions we had never encountered before. It meant we could play with genes at will: cutting, adding and exchanging fragments of DNA among surprisingly different species. We manipulated genes from bacteria to bacteria discovering and forming antibiotic tools for the benefit of many people around the world. We are composed of DNA. There are frogs, mice, chickens, sheep, dogs and cats we could begin transferring genes- plasmid- for medical and research discoveries. It does not go unnoticed of the immense use of and benefit it could bring to the food industry, including vegetables, seeds and roots. A second discovery of our planet Earth was coming as scientists were exploring micro- organisms never seen or heard of before. A new scientific language has been invented to address new frontiers in all areas of modern science. It has opened doors for women to explore and succeed in this new field of research. We have traveled a long way from scientist Marie Curie to present time. Two of my icon neuro-scientists, Nancy C. Andreasen and Jeanette Norden, are worldwide known professors whose teaching talents have motivated thousands of students on planet Earth. They are outstanding researchers and widely acclaimed University professors. We love to listen to their lectures and read their books. During all these DNA technology discoveries, it was not a single Frankenstein working behind doors in his castle in far away, Transylvania. These were men and women exchanging information by phone, journals and conference all over the world. The molecular biologists, geneticists, chemists, paleontologists, physicists as well as farmers and fishermen were turning their attention to this new technology. Viruses and bacteria were the objects of meticulous study for possible use in many areas of related investigation. Farmers the world over interested not only improving the output of crops, but also, antibiotics for their seeds and plants. Our hero, J.D. Watson was already head of the Cold Spring Harbor Laboratory on Long Island, a hot spot for enthusiastic and highly intelligent young scientists racing to make their discoveries first page news. Helping people are their main concern.

A Research Hiatus

E. Coli was a very hot tool of research not only in the U.S.A. and Europe, but also in Asia. Dr. Watson was a cautious head scientist not in a rush to open a Pandora box. There was a five year hiatus of extreme cautious research, but logic and the brain of brilliant researchers closed ranks and went on doing first class research.

Going from micro-organism to four legged relatively large animals was no small jump. Inserting the desired small fragment of DNA into a plasmid to be transferred into a bacterium, provided for the bacteria to divide, multiply and grow producing a large number of copies of DNA fragments. This step might be called the first step involving the sequencing of larger genome. And... among larger genomes, there was, in the mind of some visionary and enthusiastic scientists, the sequencing of the human genome. Once more time, the Atlantic Ocean was the divider between two brilliant pioneers in sequence technology. Wally Gilbert was at Harvard University and Fred Sanger was at Cambridge University in England. Sanger's approach to sequencing using the same enzyme that copies DNA naturally in cells, the now famous molecule DNA polymerase, proved to be more practical for its use.

Protein- enzymes functions

Proteins and enzymes functions are many. They come in different shapes located on different organs and tissues of our body. Among their functions are responsibilities for building necessary tissue for repairing occasional bruises. Another function is fighting and destroying undesirable and toxic bodily intruders like viruses and bacteria. They have done a wonderful job throughout our evolutionary history healing fractures and wounds sustained while fighting and capturing animals for food. Enzyme functions are worthy of admiration, praise and scientific careful observation and investigation. There is a wise enzyme dubbed Dicer which in another essay I called it a horseman with a sharp sword. Among its functions, this horseman cuts long double stranded RNA molecules into shorter pieces. These short pieces are crucial for gene-silencing pathways that involve small RNA such as short interfering RNA or micro RNA which are the most abundant classes of small RNA molecules."[63] It means the cutting enzyme, Dicer, the same one I had called a horseman with a sharp sword, is indispensable in generating the above named short RNA tools and their functions. A deficit of Dicer enzymes is believed to be involved in an eye disease- an advanced age -related macular degeneration. Previous work has shown that reduced Dicer levels can occur in many tissues and are associated with various diseases. Besides its crucial role in RNA- control gene silencing, mammalian Dicer1 (we belong to this group) seems to have another function. "It maintains visual health by degrading toxic RNA molecules." [64]

Scientist Hiroki Kanek, one of the leading investigators at the University of Kentucky at Lexington wrote: "Our findings elucidate a critical cell survival function for Dicer 1 by functional silencing of toxic Alu transcripts. (Alu RNAs are transcripts of Alu elements- the most abundant non- coding repetitive DNA sequences in the human genome.) This unexpected function suggests that RNAi- independent mechanism should be considered in interpreting the phenotype of system in which Dicer is deregulated. The author further alerts us by saying: "Recognition of Dicer1 unidentified function as an important controller of transcripts derived from the most abundant genome repetitive elements can illuminate new functions for RNases (enzymes) in

---

[63]- Bartel D.P., Cell 136 p. 215-233, 2009.
[64]- Nature, Vol. 471, 03/17/2011, p. 308

cytoprotective surveillance." The horse with a sharp sword I wrote about is not only involved in diseases, but is actively engaged in the most abundant non-coding repetitive DNA sequences of our genome. You recall it is the same genome sequences we used to call "junk DNA." Some congressmen were complaining the Federal government was wasting money on junk waste. H. Kanek further cautious us writing, "To our knowledge, this is the first example of how Alu could cause a human disease via direct RNA cytotoxicity rather than by inducing chromosomal DNA re-arrangements or insertion mutation through retro-transposition which have been implicated in a thalassemia, colon cancer and neuro-fibromatosis." [65]

Following signals and functions of enzymes among pathways multiple cell interventions has proven an arduous and challenging task for researchers, but it is paying off our efforts in many ways. A few years back we were complaining about the excessive amount of non-essential glia cells occupying most of our skull. These glia cells were referred to as a left over from our chimp heritage. Now we are learning they are involved in many unsuspected roles and games necessary for normal brain function. The job of micro RNAs is not fully established yet despite its wide use as a research tool, but neither is the Dicer multiforms and functions. We can see they are involved in human diseases, but to identify the loci of a particular disease among billions of cells and trillions of synapses takes time and absolute dedication. Following RNAs during their multiple functions starting with protein assembly and including short interfering fragments cutting existing double strand is a huge job. We have to add its possible participation in health and sickness demands a researcher has to contend with. When to stitch polymers together for a specialized job and when to do the cutting, even with gene fragment help, is a marvelous and ingenious intervention by any standard. Just for the sake of repetition, the job of manufacturing, degenerating and assembling units and repairing it at micro level when the need arises, is a big challenge in my book of notes. Gene splicing is

---

[65]- Nature, Vol. 471, 03/17/2011, p. 325-329.

used in recombinant DNA to insert and/or to remove genetic sequences from one organism to another DNA sequence.

Learning the trade

How gene splicing took place led scientists to explore and master the technology on how to cut and splice nucleotides sequences, meaning recombinant DNA technology. Splicing or joining together any nucleotide sequence to build a new product for commercial purpose came under the new technology. Any genetic sequence can be cut using restriction enzymes and be spliced into another sequence of letters to manufacture a medical product for diseases that have remained undefeated until now. Restriction enzymes are used in laboratories to remove or to add-splicing- nucleotides to sequences. This technology has facilitated pharmaceutical companies to manufacture large amounts of essential proteins for medical purpose. Most diabetics have benefited from this wonderful DNA medical technology. We can insert insulin producing genes into the genome of bacteria to produce large amounts of proteins necessary for insulin genesis. Some individuals may call it cloning insulin producing genes.

We will make a final comment on this most significant scientific advancement that took place during my lifetime. Perhaps we can say it took place during the last 2400 years ever since Socrates' death in Athens. During the processes discussed above, scientists discovered how their tool, a bacterium, had evolved to protect themselves from a virus infection. A wise bacterium had learned to manufacture enzymes to recognize a specific DNA sequence on the invading virus and made a cut at that threatening sequence. There is an abundance of enzymes in nature that we need to know, meaning identify and catalog. They can cut the specific sequence of a toxic invader. You just need the curiosity and enthusiasm of W. Arber to repeat his discoveries. The test tube in your laboratory is the ideal place to bring cells and extract the cutting enzymes you will need to play the game. The test tube is used to cut any DNA you want to work with. The DNA fragment that came out a cut by the enzyme can be used to splice- join together- another DNA fragment or molecule- and bingo, you just created recombinant DNA.

If during your game of cutting and splicing together small and large fragments of DNA sequences, the recipient of DNA you played with, happened to be a chromosome, you will get another bingo. You can incorporate it-meaning, introduce it into its host cell, thus receiving new genes. DNA fragments (genes) from any source can be transferred to any cell. We are all composed or made of DNA. It is the variations-mutations- within genes in the DNA molecule that made species grow differently.

There is no room for doubt if you follow the footstep of these great pioneers in DNA technology. There are many cutting nucleases in nature just waiting for you to begin a career in this field. Making cuts and splicing together all sorts of DNA will trap you forever. In my opinion, there is nothing more interesting and amusing than learning to play with your own book of life. You will learn how to engineer a bacterium to use as antibiotic vector to save sick people and return them to a healthy and happy life. You may become a genetic engineer working through worldwide agencies multiplying several times crop production by eliminating plagues, and saving millions of people from famine and subsequent death.

Viruses cannot multiply alone

You may recall that viruses by their own mechanism cannot make it. Viruses, those nasty bugs need to infect the host cell by introducing their DNA. Viruses use the cell reproduction machinery of protein synthesis to multiply themselves. A virus becomes lord and master of the invaded cell. In a short while there will not be a thing left of the host cell; viruses will be flying out looking for another victim to invade and destroy. During previous chapters of this essay you had learned how some cells have developed efficient tools for self-protection, and eventually, killing the toxic invader. We are grateful to many dedicated scientists who pursued their research goal to its end. We have named a few for you. For their contribution to science, health and humanity in general, we have no words to thank them. There are thousands of ex-patients around the world today who have benefited from DNA technology. There is a widely known and dedicated professor of Biology, Dr. David Sadava whose lectures I cannot ignore here. I love to hear him repeat things over and over again while taking real story from daily life episodes to illustrate and elucidate the message of his lesson. During one of his lectures, he described two ways or methods to get DNA into host cells. We will describe only one. You isolate a chromosome from a host organism and put it into a test tube; cut the chromosome with a restriction enzyme and the new DNA spliced- joined together- into it. This rec DNA is then put into the host cell, and since it is part of normal chromosome, the host cell thinks that it "belongs", and the rec DNA stays there and is replicated." [66]

ILLNESS UNDER THE ELECTRONIC MICROSCOPE

There are many illnesses whose etiology is just beginning to come out under the eye of electron microscope and related medical research tools. Our immune system has been under the watchful eye of scientists for many years. There are scientists dedicated to find a solution for our medical illnesses working 24/7. It has become a worldwide endeavor connecting scientists from all over the world. They are trying to unlock the door to our immune system and its failures resulting in diseases. Among them we have lupus erythematosus. This illness is closely related to our autoimmune system. It produces or provokes chronic inflammation that may involve the heart, lungs, skin, joints, kidneys, central nervous system as well as our circulatory system. Scientists in the field suspect that apoptosis is involved in the etiology of this disease. However, the abnormal production of antigens by B cells has been claimed by outstanding researchers to be the primary culprit of this illness. Like schizophrenia, autism and bipolar, among others, it presents itself as a syndrome, making it more difficult to follow its pathways in billions of cells. We have seen young people almost crippled by this disease. They become unable to walk and are limited to a wheel chair for life. Their joints, knees and fingers are deformed by protrusion- swelling- making it even more difficult to hold a fork and knife in their hands. This sad

---

[66]- David Sadava, Understanding Genetics: DNA, Genes, The Teaching Co. 2008.

condition is followed by depression and occasional suicidal ideations and attempts. I have seen young men and women praying G-d to take them away for they cannot tolerate their pain anymore.

Anxiety

My first encounter with anxiety that I am aware of took place while attending Junior High School. I had a punishing teacher, Mr. Delgo, the object of one of my most recent books titled, An Episodic Toxic Memory, a survival story. I did not have to see him or hear his voice to have my heart beat so fast and loud that even my friends used to place their hands on my chest to compare it with theirs. He had punished and embarrassed me in front of the class room. Consequently, I began to experience very embarrassing episodes like wetting my pants, vomiting, diarrhea and headaches. Mr. Delgo, my obnoxious teacher, was an ogre occupying many neurons in my brain. I can say that in some respect he stopped me from growing up emotionally and cognitively.

I had him for my teacher for three very long years. The numerous neurons and synapses involved in this episodic toxic memory were overly active-super-sensitized-firing and releasing chemicals, meaning, neuro transmitters. Ultimately, super sensitized neurons produced new dendrite sprouts. These dendrite sprouts brought about neuronal structural changes and functions in neurons. A few particular brain regions or cluster of cells were the primary recipient and reactive objects of Mr. Delgo's abusive behavior towards me. Among those cells there is a group, a very old group of cells on both sides of the temporal lobe called amygdala. It is located in front-anterior- of another group of old brain cells named hippocampus. To this primitive brain structure, people in Greece and Rome who were curious and interested in brain anatomy, it looked like a seahorse, and so they named it.

My first investigation about fear and anxiety was centered on Sigmund Freud's psycho analytic theory. My oldest brother was a fan of Freud during the late 1940's and early 1950's. I recall I had begun to feel nausea, headaches, sleeping problems, nightmares besides a racing heart that occasionally I compared it with my father's favorite horse, Canelo. I tried over the counter medication and herbal teas. My mother was believed to be an expert botanist among her immediate family and friends. In all honesty, neither my mother's expertise nor o.t.c. medication were of much help. My problem was not in my stomach despite my nausea and occasional vomiting. My problem resided inside my head somewhere in an outgrowth of synaptic connections I could not even imagine it existed at all. Eric Kandel, Jeanette Norden and Nancy Andreasen, three of my earliest icons and intellectual stars, helped me get insight into the monster I had right inside my brain torturing me long after I had left Mr. Delgo behind. I had left behind Delgo in bones, blood and flesh, but his image, his bullying and all his wicked behavior was inside my brain. Later on, when I had moved to live in a great urban area with many experts in brain-mind-problems, I spent almost three years on a couch followed by another three and half years at a psychoanalytic oriented institute. The institute had gone significant modification from traditional Freudian theory. I must add that group therapy and group support were extremely helpful.

An almond looking cluster of brain cell

The almond looking brain structure we are referring to is the amygdala, which along with the hippocampus or seahorse, is involved in memory and learning. Without that almond looking structure positioned anterior to the hippocampus, we would not probably have made it in the jungle. I am not striking against animals living in the jungle, please. The new predator drives or rides expensive cars, wears expensive clothes, makes use of most modern cell phones and hires less sophisticated predators to do his job on street corners or expensive and selective bars.

I like to travel and my money has vanished from my pocket leaving me without a minimal of suspiciousness about who was the culprit or thief. You may argue that the amygdala is losing its usefulness, but perhaps it is the predator that has become smarter. The amygdala acts independently of nearby clusters of brain cell nuclei. You have the limbic system and prefrontal cortex that are emotion power centers. However, the amygdala does not wait for feedback or response from them when it needs to put you in a safe place in a fraction of a second. Otherwise, you become lunch or supper for a hungry predator. This amygdala of yours is often referred to as the brain door to emotions. There is a difference between fear and anxiety. In fear there is an object you are afraid of. In my case, I began to fear Mr. Delgo while I was in his class room. Later on, when I had moved to live miles away, fear was converted into anxiety. Anxiety has to be defined as a sustained level of elevated apprehension in the absence of an immediate threat. I had moved 1600 miles away from Delgo, my teacher, but the ogre in my brain kept torturing me day and night. He could not be an immediate threat to me at that distance, but he had already provoked a structural change on my neurons. When I saw a man who looked or spoke like him, anxiety overcame my apparent peaceful mind. When I was enjoying an ice cream sugar cone and a Mr. Delgo look alike passed by me, the ice cream turned sour and occasionally, made me throw up. Anxiety is considered by many clinicians as the most common psychiatric disorder. It may come from different sources and it exists in all cultures and countries around the world. It may go unrecognized, misdiagnosed and untreated for many years in all levels of our society. Denial and the traditional stigma attached to individuals receiving help or treatment for brain problems keep people to stay away from receiving help. Unfortunately, they remain in hiding rather than getting professional help and regain their happiness. It goes without saying that alcoholism, drug abuse, marital discord and abuse have their base in anxiety.

Our subject of discussion, the amygdala, has been implicated in anxiety etiology. Its implication is logical and is based on many laboratory experiments. Chronic stress can kill many cells in the amygdala, thus contributing to anxiety. Dead cells in the amygdala make you forgetful, learning is slower and you become irritated. Your blood pressure goes up, headaches are turned on and strokes and heart problem may also come to make you feel sick. Most medical doctor`s visits have an anxiety component. My story on the amygdala and anxiety is not intended to frighten you. On the contrary, my good intention is to help you understand your emotions. It is best to familiarize yourself with relative minor problems that can be resolved before it reaches the brain and makes it its permanent home. Despite the high prevalence of anxiety disorders in all levels of society, the underlying neural circuitry is incompletely understood, according to a Stanford University professor and researcher. Psychotherapy, among them, Cognitive Behavior Therapy, Gestalt Psychology, Rational Emotive Therapy, and goal specific and action oriented therapeutic modalities have been very helpful dealing with anxiety disorders and depression. There are prescription drugs like the benzodiazepines, but their side effects and possible addicted component need to be seriously considered before you begin to take it.

In addition to incomplete studies on the amygdala, this investigation deficit can be partially attributed to the complex internal composition of the amygdala itself. The leading author at Stanford, Kay M. Tye wrote that the amygdala is composed of functionally and morphologically heterogeneous sub-nuclei with complex interconnectivity. The primary neurotransmitters are glutamate and GABA ergic medium spinal neurons. The centro-medial amygdala consisted of 95 percent GABA ergic while the basolateral amygdala is glutamatergic. No conclusive evidence of a specific causal factor or loci for general anxiety at the amygdala was found. Their investigation concluded that anxiety is continuously regulated by balanced antagonistic pathways within the amygdala. [67] During this essay as well as previous ones, we have paid particular attention to the amygdala and the hippocampus not only because its antiquity in the brain; but also, because they are always on the hot spot. Both are primary survival cluster of brain cells that have helped this lad reach this point in life's evolution throughout thousands of years. Soon, we will identify a particular protein in the amygdala responsible for breaking the established normal balance between glutamate and GABA production, thus provoking an emotional disorder. It is the amygdala's continuous exposure and over-excitation to stress that governs that particular protein. Likewise, we picked on translation

---

[67]- Nature, Vol. 471, 03/17/2011, p. 358-362

processes where RNA is always a big player, but also, a hot spot for human illness. We should not close up and say, hasta la vista, until we pick another hot spot: cutting and splicing, particularly, splicing involved in many of our illnesses. Some researchers have begun to explore that sensitive area of cell manufacturing machinery. It would be wonderful for all of us if we could address this health issue with embryonic stem cell lines and re-generate a decaying amygdala. It does not escape my mind-brain that progress we have made with an engineered bacterium can well be an excellent vector to make appropriate corrections and adjustments in a dysfunctional amygdala.

AN ANIMAL CELL AND AMINO ACIDS IN PICTURES
Table showing the 20 amino acids and stop codons

We will write them beginning with the largest number of codons to the smallest number of codons.

Amino acids   Codons

Arginine-----------------AGA---AGG---CGA---CGC---CGG---CGU
Serine--------------------UCA---AGU---AGC---UCC---UCG----UCU
Leucine------------------UUA---UUG---CUA---CUG---CUC---CUU
Threonine---------------ACA---ACC---ACG---ACU
Valine-------------------GUA---GUC---GUG---GUU
Glycine------------------GGA---GGC---GGU---GGG
Alanine------------------GCA---GCC---GCG---GCU
Proline------------------CCA---CCC---CCG---CCU
Isoleucine---------------AUA---AUC---AUU
Cysteine----------------UGC---UGU
Phenylalanine---------- UUC---UUU
Tyrosine-----------------UAC---UAU
Histidine-------------- --CAC---CAU
Glutamatic Acid------ -GAA---GAG
Aspartic Acid-----------GAC---GAU
Asparagine--------------AAC---AAU
Glutamine--------------CAA---CAG
Lysine----------------- --AAA---AAG
Methionine----------- --AUG
Tryptophan-------------UGG

Stop Codon--------------UAA---UAG---UGA

     We arranged it from the largest to the smallest number of codons to code for protein synthesis responding to our curiosity and convenience. No relationship between each amino acid and its codons number from another amino acid was intended.

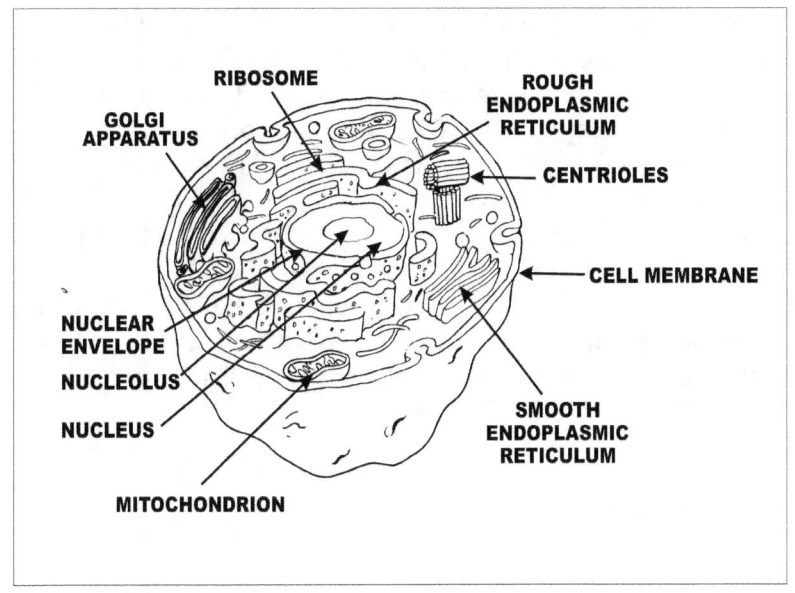

Inside an animal cell

There are dynamic chemical and molecular processes taking place inside each cell every fraction of a second of our life. Most of these processes involve extremely tiny units seen only under an electronic microscope. Equally small there are multiple vessels or organelles traveling back and forth inside the cell plasma carrying and delivering organic material necessary in our organism. You will find the endoplasmic reticulum, the Golgi apparatus, lysosomes and neighboring cells.

For instance, a tiny portion of the plasma membrane on the surface of the cell may separate and cut itself from the membrane, capture necessary neighboring subunits and bring it inside the cell for processing. Similarly, these tiny cut off portions may be used as carriers to nearby cells. There are always intercellular captures of material going on for cell processing and use. Captured external material is usually fused with lysosomes. These cellular vessels, lysosomes, function as a de-generator machine digesting everything that passes by. Part of this dynamic process is importing and exporting cell material. The nucleus of the cell is surrounded and protected by a membrane often called, nucleus envelope. Right next to the nucleus, you will see the endoplasmic reticulum that is always busy processing and packing material inside the cell. Keep your eyes wide open because within this particular cell vesicle you will find the protein assembly factory, the ribosome. In each eukaryote cell, meaning cells with a nucleus, there are many compartments with specific roles. The aqueous material- gel like substance- is called cytosol. Besides serving as a cushion, many cellular actions and reactions take place there. We must remind you that inside the cell there is the energy producing molecule, the mitochondria. Please, take time to look for the nucleolus. You may call it RNA's home.

FROM RESEARCH TO CLINICAL PRACTICE

It was a hot and humid mid-July 2011 evening in Brooklyn, New York City. The Earthquake and Irene, the hurricane, had made an unwelcomed visit to us. I had already done my trip around the world in the internet searching for outstanding news and discoveries in science, economy, and politics that could have an impact on my life. For scientific discoveries, I look for and rely upon the most prestigious journals and books published around the world. My main concern about world politics and economy is the Middle East, Russia, Europe, and of course, the world's biggest factory, China. It is not that I ignore Africa and South America, especially, Brazil. This huge country with immense resources and recently discovered off-shore oil deposits will influence world economics and politics in a fashion unpredicted a few years ago. China and Russia are opening doors in South America that may soon become the most competitive world region besides the Far East. It is not just agricultural genetic engineering, lumber, beef production, but also, home grown and imported scientific technology that is transforming the sleeping giant into a world power with its own particular interest and agenda. These issues keep my brain active making relevant associations and connections between dendrites. This mental exercise helps strengthening synapses making new memories that may become permanent in my brain. The old saying of use it or lose it, I do not ignore or put it away. It keeps my brain from becoming rusty, just as my grandfather used to say.

At home, my bed-room is full of books and journals, mainly, on neuroscience, psychology and genetics. I do not do much exercise because of my breathing problems; I developed emphysema. I did not order emphysema online, it just came into my lungs without an invitation. I take medication on a regular basis. I walk a few blocks every day, and I am choosy on what I eat and drink. I call myself a happy and blessed person. People stop me for all types of questions including health and world politics. My engagement with them is not linear or one way dialogue, but it takes many directions often answering a question with another question. Many times I tell them that if I answer their questions, I will deprive them of the joy of finding the truth by themselves. This is done based on my conviction after many years of clinical practice. Repeating somebody else's answers, whether true or false is robot like behavior. I am not saying that you do not need to learn from your neighbor. What I am saying is that each one of us is a unique individual unlike anybody else in the universe. Whatever truth you are seeking out, it is your own truth, not your neighbor's truth. Even identical twins do not share the same truth. Each one of us has a genetic code with information many hundred years old. Besides, molecular and chemical reactions taken place, above all, in our brains, makes you different from anybody else here and now. The brain avoids many problems and diseases if it is busy working on something that will benefit us all. My daily chat with people is a form of socialization that is helpful to me and others. There are people twenty or more years younger than I am who keep asking me for my recipe, meaning my daily diet. Besides my local chats, I keep in contact with friends and colleagues through the internet or cell phone. After I retired from my clinical practice after my 79th birthday, I have a bad habit of visiting my friends at home or at their office. They have all responded happily to my visits. I write up all this for you because I want to share with you a recent unhappy experience that is bugging my brain and I need your input.

On that hot and humid mid-July evening, actually, it was 11:00 P.M when I decided to call it a day and go to sleep. Usually, before I turn off the lights in my room by pulling a yellow cord located next to my left side of my head. I confess that I have a bad sleeping habit of grabbing a book, journal or magazine for my last light reading for the night. Believe it or not, but the purpose of this last reading is to relax me into sleep. Purposely, I pick up something that would not be provoking to my brain. I do not want to read that late at night a scientific discovery, new medical discovery or the latest DNA technology that may excite my brain cells into further investigation. My last few minutes before my brain sleep-awake circuit is turned off, have to be spent reading innocent and soothing stories. Unfortunately, I do not find those pleasant and sleep inducing pages all the times. And…this mid-July hot night was one unfortunate and miserable night experience for me. Do not get me wrong, please. This is not the only painful and torturing night for me. I have had more than a just share of miserable and torturing night experiences during my life time. However, most of the time, it served me well as a good learning experience. I bet that either you or I will spend 83 years of life on planet Earth just listening to old sweet children stories. I love those stories, but there is a time and place for it. I was not born on a golden crib and I attended public overcrowded schools. I am just like hundreds of millions of children, adolescents and adults who grew up in sweat, hunger and unjustified rejection and punishment. It becomes an integral parts of our life's trip in this beautiful home-planet of ours. I am not implying that growing up is an easy task for a human embryo or any other organism, but for many of us, the rules of nature have been heavy handed. My inquisitor interrupted me asking to stop complaining about poverty, disease, hunger and cruelty of all sorts. He added that the world's standard of living for human beings has never been as good as it is now. I would not argue with him that point. I myself have had pleasant, beautiful and educational experiences that make me appreciate even more the beauty and goodness of this planet and its people. I try to enjoy it as much as possible.

Before you put me away, I am going to tell you what struck my brain like a lightening rod that unforgettable mid-July night. There was hardly any cluster of brain cells that did not lit up and provoked unwanted behavior on me that I was not aware of for many, many years. My brain has a habit of going traveling in space and time even against my will. It takes me to places and time I have never imagined it existed even in fairy tales. And, believe me, I have read many fairy tales beginning with Don Quijote de la Mancha. Before I forget it, I am going to tell you a secret I did not reveal even to my mother. Once I dreamed I could write a book like Don Quijote. It is not a joke; I dreamed such a non sense. Go ahead and have a good laugh at my expenses, but it is true, I had such a contempt for Cervantes that even unconsciously (dreaming) I made fun of him. Responding to my ever torturing inquisitor, I am going to cut short this preamble to the experience that drove me almost insane that otherwise quiet and peaceful night. You can call me a bleeding heart, a coward, a panicky old man, or a messenger from heavens, and I will forgive you. Forgiveness, as I see it, is a powerful healing tool. Anger or jealousy, on the other hand, are as poisonous to your health as is a cancer that eats away cell by cell before it is discovered. We have the tendency to deny our anger while it destroys the inside of us. Our brain suffers, our heart suffers, and our stomach develops ulcers if we do not become honest with ourselves. We spend a life time defending our own lies. We behave even worse when we buy all types of remedies to cover up our ulcer and headache. If you allow your organism to take care itself without your un-necessary interference, you will be better off.

I have to implore your mercy not to be over exaggerated condemning my emotional response that horrible evening. The motive that triggered a non- anticipated, spontaneous, unconscious and automatic affective response on me that night, made me revisit in its entirety, my childhood life experience. Be prepared for what is coming; do not hold back your emotional response. Remember, the first thing a baby does when he or she comes out of the comfort inside the mother's womb, is crying. The visual trigger that provoked my emotional make up was a girl about six or seven years old with a strapping above her left ankle tied up with a metal chain to a tree trunk. Repeat, the girl was chained to a tree trunk. It was shown in large scale at the beginning of a highly professional and prestigious scientific journal. I turned on a second light to make sure that I was not exaggerating nor my brain was creating a false design from the picture I had in front of me. Please, do not stop reading me; try to get the whole picture before making any judgment. My intention that night, although naive, was to get a restful and enjoyable full night sleep. That was my mother's last blessing at night before I plunged into deep sleep.

Right below the picture, the article's author explained that the girl was punished by her family and neighbors because she was claiming to have seen and talked with people not visible to others. It is hard for me to believe that you have not heard a story like this one before. In our daily parlance, the girl was having visual and auditory hallucinations. Some cultures may claim the girl was blessed from heavens for the good of the many. However, in this case, the girl was punished because her family and her community believed she was bewitched by evil spirits. They believed she would bring diseases and death to her neighbors. The picture and the story about this young girl were published for public view during the year 2011, not four or five hundred years ago. It was taking place in our time and in our planet. It was not a television show or documentary aimed at a specific audience for commercial purpose. Incidentally, on July 18, 2011, a public financed television channel aired a documentary about a southeastern Asian country where schizophrenic patients were kept in bamboo made cages. They were considered a threat to society. A shaman, not a mental health professional was responsible for the diagnosis and treatment of brain problems and diseases. This brain disease is not limited to a country, race, social or economic class; but diagnosis and treatment in underdeveloped countries are extremely poor and inadequate. Industrialized and relatively rich western countries have well trained professionals and medicine available to control most of the symptoms of this disease. Most people on planet Earth do not know that genes and biochemical markers in brain cells may be responsible for the voices the poor girl was experiencing.

My behavior during this episode that night of mid-July, I am still trying to break down into its component pieces for proper analyses and digestion. What is to be assimilated, if it needs to be, is too early to tell. Before I could place the journal and the picture aside, I noticed that tears were running wild on both my cheeks. Also, my nose was getting wet. I have worked with children, adolescent and adults for many years. My experience at a psychiatric emergency room placed me in contact with patients whose problems broke down even the best trained and matured therapists. Whether you believe it or not, but the pain of others, sooner or later begin to wear you down. It becomes a revolving door; therapists left the job no matter how comfortable we tried to make it for them. Yes, that hot night I found myself crying for a little girl I had never seen. Whether the tears were over the chained little girl, needs to be further clarified, explored and analyzed. We carry emotional scars in the form of toxic episodic memories that may be triggered by the pain of other human beings. Even the abuse of kittens, puppies, chicks or any other defenseless animal may serve as an emotional trigger for a physical response.

My eyes saw a girl in pain tied up to a tree trunk; the message was sent to my visual cortex in the back of my head. From here, it must have gone to the thalamus, a relay stationed composed of many complicated brain cells deep in the middle of my brain. The limbic system on and around the thalamus was immediately involved including the pre-frontal cortex. Groups of brain cells included in the limbic system seem to be an arbitrary one. There is not a consensus among neuro scientists over which brain cells nuclei should be included in this system. But most agree that it is a very primitive survival region of the brain and seat of emotions. And…when we talk about survival, the amygdala takes a front seat. The amygdala has not only helped me survive my 82 years of age, but it drags within itself automatic responses from thousands of years of human evolution. We will come back to the limbic system later on in this essay. I hope I have already named most clusters of neurons included in this system. I am careful not to contaminate my brain with dangerous drugs, toxic memories and grudge against anyone. When you hold a grudge against another human being, you are hurting yourself. You have turned anger inward and it manifests itself in many bodily symptoms you try to cover up. You drink or swallow all sorts of things calming down your symptoms; but in reality, what you are doing is hiding your lies from yourself.

For my tears to flow freely against my conscious awareness, they must have touched deep brain pain spots that might have been formed years ago. They have become permanent memories stored in my brain in a place I cannot identify at present time. Forgive me for the interruption, but next to me is sitting my old friend. He keeps bugging me with his opportune questions.

Yes, Mr. Inquisitor, the hippocampus was involved in retrieving past memories. You must be familiar with the accidental case of H.M. who could not remember anything after surgical removal of each hippocampus in each temporal lobe. The H. M. tragic case and most recent studies have confirmed without any doubt the function of the hippocampus in forming and retrieving memories. Anterior to, and very close to the hippocampus, we find the amygdala that is strongly involved in learned fear. The big tragedy in H.M.'s case consisted that both hippocampus were removed. Each temporal lobe has one. H.M. suffered from chronic and very painful epileptic seizures. PET scanning and functional fMRI, two sophisticated technologies, have been very helpful during the diagnosis of brain diseases. In vivo imaging techniques while working with patients in clinics and hospitals give scientists a very clear and accurate picture when everything in the brain is working well. Likewise, an abnormality in structure, function or damage to a brain cluster of cells will be obvious to the expert eye of the scientist doing the job. The above clarification is in response to my inquisitor who does not miss a chance to interrupt me. The truth is that I benefit from his occasional observations and comments. I hope you too forgive him. I must return to the little girl chained to a tree trunk before he provokes me again.

I said the picture struck me like a lightening rod, but in retrospect, I must modify the term because I have not seen or heard anybody survive such a punishment from the sky. Besides, I am trying to find connections between the lightening rod, the metal chain, the girl and my bed. As a country boy, the first thing that comes to my mind is lightening rods hitting palm trees and avocado trees and splitting it in pieces. That is a very frightening and dangerous experience for anyone, young or adult. I will leave lightening rods alone and address my fear and tears on my cheeks. The first sight of the girl looking down, and apparently, in tears must have stricken a group of my brain cells containing very old and painful memories of at least 75 years ago. When I was about seven years old I saw an old lady with her hands and feet tied up together taken to a hospital because she had put in flames her family home. Later on, as an adult, I learned she had a schizophrenic daughter and a very violent son. During my private practice, I worked with a patient whose depressed mother tried to kill him at age 5 by tying him to a tree trunk and setting on fire a nearby ranch. There are painful memories in my brain that could have been triggered to action by the sight of a chained little girl. However, it seems that my first unconscious protective response was to ignore what I had seen and turn on the next page. What I had already seen, but did not want to see, must have been associated with or linked with hidden memories repressed by defensive layers of denials. Denials and repression may help you momentarily, but in the long run, they will catch up and get you when least expected. When I had already turned my attention on the next page trying to deny what I had already seen, a slight feeling of anger showed up on me. Quickly and without hesitation, I went back to the little girl in chains I was trying to ignore. I turned on a second light in my room and it convinced me that my brain was not playing tricks on me with unconnected designs and visual patterns. Not totally convinced that I was reading something taking place in our beautiful planet during the 21 st century, I took another look at the cover page. My brain was not playing tricks on me; the date in front of the page read: July, 2011. I went back to the page with the chained girl and looked very closely at the chain. I was looking for something to soften the girl's pain, which in reality, it was my pain. I was trying to change the metal chain for something less painful and ugly to my eyes. I was trying to put humanity to the beast that had chained an innocent human being to save his/her own ignorance and wicked behavior. No matter how hard I tried, the beast was over powering, and the chain got heavier and brutal every time I looked at it. I saw men, women and children all over the world in chains. I could not help it, but my brain took over and sent me in time and place traveling to Egypt, Persia, Babylon, Greece, Rome, Africa and America. It is still talking place in the 21st century. We punish people for things we do not understand.

I saw nothing but chains, chains covered by blood of innocent men, women and children. It looked as if a beast like creature was feeding on human suffering and blood. However, after I attempted to uncover the beast what I found was man's thirst for power and greed. The beast is a man made animal; it enjoys torturing and killing other human beings. The main focus had been the girl suffering in pain because she had shared with family members and friends things she alone had seen and heard. She might have been asking for support among trusting family members. My brain tried to soften up my own pain by generalizing pain on other human beings and took me on a mind world trip. I have to be grateful and thank my brain for trying to ease my pain; it has done it for many years. It may also be interpreted as an unconscious attempt to find an excuse, a rationalization to continue to deny the little girl had pushed pain buttons in my brain. Perhaps it was true; the chained little girl had exposed my weakness. Despite my intense training and clinical experience there are scars that are not completely healed. My attention was divided into two different objects: the little girl and the chain. When I realized my brain was easing my hidden pain by generalizing pain in many people around the world, I went back to the innocent and defenseless victim. It was then that I could see, almost hidden between different shades of color in the picture, tears coming out of her eyes. You must remember that I was not aware I was crying. My tears were running down my cheeks and my nose was getting wet, but I kept looking at the picture as if it were incomplete. I kept on looking at the metal chain wondering if it could have been made of softer material. A soft material could have made the pain, perhaps more tolerable. Actually, I had taken my eyes off the triggering stimulus that provoked my own tears: the little girl tears. My brain world trip could have softened my own pain and put me to sleep, thus avoiding going back to the visual trigger that had provoked my silent tears. My brain wanted to spare me from further pain and chose the chain as an attractive object to re-direct my attention.

Tears by themselves can be triggered by different sources. Some people shed tears of joy when they meet family members or receive good news from neighbors and relatives. I have seen men and women shed tears when their baseball team made it to the top. And I have seen people crying over the death of a little puppy or kitten. The stimulus that triggers the shedding of tears seems to be a complicated issue. Professor Antonio Damasio is head of the Department of Neurology at Iowa University Medical Center. In his book, "Looking for Spinoza", he describes for us an extremely interesting episode of tears shedding that relates to my case. It is a real case documentation of a 67 year old lady with a history of

Parkinson disease no longer responding to conventional medical treatment. She was a candidate for deep brain stimulation that required implanting two tiny electrodes in the brain stem. The brain stem is a primitive or very old region of the brain that begins in the upper part of the spinal cord. Generally, brain anatomists divide the brain stem in three sections: medulla, pons and midbrain. It is a survival region of the brain because it controls many vital organs of the body. Surgery for implanting the electrodes took place in a hospital in Paris, France. The patient, Jeanne, not her real name, had no history of depression or anyone in her immediate family. She was one of about 19 patients who had undergone the same procedure at the hospital with the same surgical team. Each tiny electrode has equally tiny points of contact to deliver controlled electrical current. By misfortune, the electrical current missed the target by a tiny fraction of two millimeters and touched what Dr. Damasio called an emotional button: sadness. Otherwise, she was responding extremely well to surgery. Now listen carefully to this part of the story without pre-judging anything. To the surprise of the medical team, Jeanne, the patient, began to cry shedding tears and sobbing loudly. She even verbalized to the staff she wanted to terminate her life, meaning she did not want to live anymore. She was calling death to come and take her away. She had gone for surgery with the intention of living and enjoying a few more years on this beautiful home planet of ours. However, a little mistake in her brain changed her mood from happiness to a death wish. This episode is translated into an electric current turning on a trigger point in her brain that provoked a depressive state with suicidal ideation. There was no history of depression in Jeanne's life history. As far as the medical team was aware of, Jeanne was a happy go lucky person, except for Parkinson disease. I brought this case to you because I have been talking of pushing pain buttons we are not aware of. It is another confirmation of the existence of pain buttons in our brain. In this case, the action of pushing the button on Jeanne was done by an electric current. Jeanne was not complaining of having pain because the electrodes were hurting her; the brain per se has no pain receptor cells. Your body skin from head to toes is responsible for those receptors. After the electrode position was corrected, Jeanne made not a single sob or shed tears. Her death wishes disappeared from her like magic. She became the same jovial, talkative and happy lady before surgery. An interesting observation Dr. Damasio does not want to go unnoticed by us is that the touching of the emotion: sadness, came first, and feelings, a subjective state came afterwards. In my case with the little girl shedding tears while chained to a tree trunk is still unresolved. We cannot blame the electrodes or anything else in the surgical procedure as the causal agent. I was alone and comfortably lying down on my bed. The messenger that carried the information that provoked my tears and, consequently, the feeling of guilt and anger were my two eyes. How did the message go to a specific pain button that late in the evening when I was getting ready to set aside the day's challenges and go to sleep, is another question to answer. I had eaten a good dinner on around 6:00 P. M. and I was rested before I decided to say good night to everything around me. Unlike Jeanne, I did not have anyone working on my brain or asking me provocative questions. I can say that I was happy that evening; no intrusive thought or thoughts were breaking a harmonious chemical balance in my brain.

Well, I have to admit that I was crying over a picture of a girl I had never seen before. But there is another angle in this picture I have not addressed it yet. This one is not an intellectual exercise, but one that sits at the base of my emotional response to this painful experience. How come I responded with anger when I realized I was running away from the real issue at hand: tears running wildly on both cheek. Why anger, and where did it come from? How tears and anger come together, if it comes at all. How is the little boy in me involved in these emotions? Do I have to go back to my childhood life experience to try to explain my adult tears and anger over a picture of a little girl crying while tied up to a tree trunk? We can make use of the Freudian psychoanalytic theory to try to explain, both my tears and my anger. We can call on the ego as the main agent delivering messages and mediating psychic conflicts between the id and the superego. However, I prefer to stay in the 21 st century and look for neuroscience to explain my behavior based on scientific proven models. There appear to be significant life events between Jeanne and I. She did not suffer from depression and was a happy lady all her life. I, for one, did not enjoy a happy childhood. My mother had eleven children and, my brother born right before me, suffered from what appeared to be thalassemia major and died when he was a little over two years old. My mother had confided my care to my oldest sister. Secondly, while attending intermediate school, meaning, during the 7th, 8th and 9th grades, a teacher named Delgo provoked a very painful, traumatic and toxic memory on my personality. See, An Episodic Toxic Memory, a Survival Story by this author for details. I spent years in therapy trying to bring closure to a dysthymic disorder. It can be called a mild version of a depression. Psychoanalytic oriented therapists would not hesitate to connect my tears to my childhood experience, but I prefer to chat with modern science and technology.

In Jeanne's case, Dr Damasio does not tell us the specific loci, meaning the location in the brain stem the electrode that provoked tears and sadness were implanted on. That section of the brain has many nuclei or cluster of brain cells responsible for myriad of behavioral response. Besides, the brain stem appears to be similar to a super highway, extremely busy with messages going up and down. Messages from the body go up the spinal cord to the cortex, and likewise, the cortex sends down the rest of the organism messages through the brain stem. Adding to the complexity of this region of the brain, it has clusters of cells responsible for generating chemical messages like serotonin and nor-epinephrine responsible for behavior and memories formation among its functions. There is another group of cells around the medial superior part of the pons, not far from the locus coeruleus and Ralph nucleus known as substantia nigra compacta. When its cells die out, it no longer produces dopamine. When there is lack of dopamine from this brain nucleus, the person develops Parkinson disease. This is the illness Jeanne was suffering from and electrodes were implanted on to make up for the deficit of dopamine. The dopamine released from the substantia nigra, go to two nearby cells nuclei known as caudate and putamen. Dopamine is not only involved in Parkinson disease, but also, in schizophrenia. I am not trying to get you involved in brain anatomy and physiology. It would be wonderful if you do, but what I am trying to say is that it is not easy for me to say: "here is the solution of the problem for you." People laugh, cry and shed tears for many reasons they themselves cannot explain. In a certain way, that is exactly what I am doing now. My first reaction was to run away from the girl's tears, but I returned to her in anger. Anatomically, both, tears sites and anger are not far from each other. Tears come out of your eyes and are visible to an observer, but the button location you already saw in Jeanne was visible only to the surgeon through sophisticated scanning machines. It was a tiny metal that missed its target. Anger, another emotion, has many ways of manifesting itself, but the locus in the brain is not far from tears. The hypothalamus, deep in the center of the brain, but not part of the neo-cortex that developed later on, is involved in anger formation among many of its functions. The hypothalamus and the pituitary gland, a crucial endocrine gland is structurally and functionally like identical twin brothers, inseparably behavior. There are hormones traveling from a cell nucleus to another carrying a message that can be translated into anger. For the sake of clarification, let's say that Pete responds with anger to Mike's joke while John just laughs at it. Mike might have taken a stone and hit a strayed dog on the street. Pete might have passed from an anger phase and engaged in a fight with Mike over a joke and the hurting dog. In this hypothetical case you have two potential anger provoking sources: visual and auditory. John remained neutral. Pete's brain was responding to two old and powerful sensorial input. You may point out that Pete parents' religion and cultural background must have contributed to his behavior. Pete's upbringing might have been different from that of Mike and John. Pete also may be responding to Mike's bullying and aggressive behavior. Mike hitting a strayed dog burst opened the anger bubble. A section or nucleus of the brain had been accumulating painful experiences and feelings that no longer could tolerate. In Jeanne's case, her tears and suicidal ideations were not provoked by a religious, cultural or car accident. A piece of wire had mistakenly touched a tiny piece of brain tissue. Instead of touching an anger button, the wire came in contact with a tear shedding and depression provoking loci. Tears and anger have multiple functions. Tears can be lubricants for our eyes cleaning and protection. Anger, besides its function in aggression can thrust you into battle in the defense of your country and family. During the research section of this essay, I took time to expose a myriad of cells responses and behavior. With that in mind, please try to

174

understand how difficult it is for me to explain my emotional response to the little girl in chains. Returning to my experience, unlike Jeanne and Pete, there were no electrodes or bullying person like Mike to provoke tears shedding. It was a picture of an innocent little girl chained to a tree trunk. She was not hurting anyone. She was attempting to get help from family and neighbors to explain something she could not understand. In Jeanne's case, a professional health team was helping her overcome a brain deficit in dopamine and re-establish normality in her life. We are assuming that Jeanne was not verbally assaulted nor she heard anyone screaming while she was undergoing surgery. In other words, she was not receiving visual and auditory painful messages, meaning stimuli. She did not have Mike to bully her. In my case, I was bullied during childhood and as adolescent too. Can these past experience over 65 years ago have anything to do with my shedding tears? If I am not mistaken, most people would dismiss it as nonsense. They would argue that it must have been forgotten long ago. It seems most logical to most people that time would delete past bad memories. Our culture encourages us to ignore or to play down childhood experiences as significant during adult life. We are told the same song over and over again by peers and family members. People around us usually say: You are man or a lady now; forget childhood stories We are adult now, we should not be entertaining unimportant nuisance from earlier years. Boys are encouraged to hold on and avoid crying while girls can shed as many tears as they want to. We can call on Doctor Antonio Damasio, he rightfully reminds us that we are an emotional being. During my own clinical experience as well as psychodynamic theory and practice, I dealt with emotions in many details. We believe that it is a strong point to attribute anger as well as the emotion sadness to a biological base in the case of Jeanne. The fact that she responded to a touch of a tiny metal by shedding tears and crying without any other provocation amply supports our thesis: humans are basically, emotional beings. We are conditioned to repress feelings at the expense of our mental and physical health. We have on each side of our brain, in the temporal lobes, a master nucleus of emotions called the amygdala. This very old cluster of brain cells have helped us survive many years of jungle life. We have carried the jungle with us to urban areas where violence against each other is a day by day survival experience. And, of course, our amygdala has made appropriate adjustments to meet the demands of traffic jams, job competition and insecurity, street gangs and family needs, among many others. The dynamics that triggered my tears is at present time unresolved. What provoked me to look deeper into my emotional make up are multiple causal factors. The tears came out spontaneously as if it were pushed out by an internal mechanism out of my conscious control. I was not sobbing or screaming out of pain, but my emotional me was physically responding to a stimulus. It was a visual stimulus that called for a response involving tears shedding. Was the emotion sadness involved in this case? Please, remember, Jeanne was shedding tears and verbalizing a desire to finish her life minutes before. She had been a jovial and happy patient until the electrode touched her. She caught the surgery team by surprise and emotionally unprepared to give an appropriate response when she verbalized death wishes. I responded to my tears by going to the next page of the magazine. I could argue that my unconscious defense system is intact and well prepared to handle unwanted emotional situations and challenges. I can formulate this scenario in this way to facilitate my point. My defense system experienced my pain and ordered my motor cortex responsible for moving my hands and fingers to change pages in the magazine. By changing pages in the magazine, I would not only stop looking at a painful stimulus, but change to a pleasant and less provoking stimulus. It might have changed my emotional state at the time. I would have stopped shedding tears and a wet nose. That would be similar to the amygdala telling me in case of danger: legs run as fast as you can before a mad bull gores you up. Whether this is a poor comparison is a different theme, but I hope you get the point I am trying to get across to you.

When I became aware that I was running away from the challenge the picture of the little girl was posing on me, I became angry, angry at myself. I was not asleep or sedated for me to go into the next page unconsciously. My defense system was protecting me from shedding more tears. It was turning off or blocking unconscious pain receptors. This attempt to find a solution to my shedding tears, and then feeling angry is not too uncommon. When I was running from a mad bull in my father's farm looking for a safe place, if I got bruises or was stung by a crazy honey bee, my defense system would take care of it and spare me the pain. After I had gotten to a safe place, the pain would appear. You and I have built an organism throughout our evolutionary journey that has learned many tricks to take care itself. Blood clotting is a very useful and protective tool that prevents us from bleeding to death. The feeling of pain is another trick your body developed to let you know there is something wrong going on in your organism. Even inflammation under our skin is a protective device; it means your immune system is working. It means your T cells and legions of its assistance cells are doing their job. They are placing the invading pathogens in a corral for the final kill. Even tears have a protective role on your eyes. No, no, I am not running away from the main question at hand. I am supporting the thesis that both, my tears and anger could be defense mechanisms developed by my organism to spare me from additional unnecessary pain.

During Jeanne's surgery and unintended provoked depression symptoms, the removal of the electrode tiny metals stopped all her symptoms. In my case, my organism past experience with pain must have placed into action a stop signal at a gene or protein level. The visual sensory stimulus had triggered, meaning turned on a gene or genes for the synthesis of proteins that would mobilize a whole mechanism for tear shedding. I hope that you agree with me that we are not conscious of action taken at cellular, molecular and atomic level. I do not know what my genes are doing as I type these words. I have a conceptual knowledge about my brain's many actions taken while I do the thinking and type what I am thinking. However, I am not aware of all the things my brain is accomplishing. Can you guess how many chemical enzyme reactions have been provoked to do the things I am writing right now? Computers can come out with a good number; but still an educated guess of what really happened inside my brain. I am not trying to make things more complicated than they already are. Tears and anger provoked by a photograph of a girl I do not know of or seen before, is complicated enough for me to add anything else. First, I have not seen anybody torturing an innocent and defenseless child in that fashion; except the story of a five years old boy almost killed by his mother during a psychotic episode.

Actually, it was a picture on a piece of paper that provoked both physiological responses on my part. As far as I know, I was not suffering from a disease that might have provoked unwanted tears and anger. I had gone for a complete physical examination including blood work. I said earlier that I did not want to get psychodynamics involved in the case. Likewise, for readers with a strong religious orientation, the picture of Christ and Mary may provoke tears. During my visits to Fatima in Portugal and Guadalupe's shrine in Mexico, I have seen many people cry and shed tears. It would be a complete different route if I pursue my tears experience under psychodynamics or religious interpretation. I am trying to see it from the perspective of an organism that is rational, persevering and emotionally aware of its complexity. This organism believes that emotions are at the base of humanity. It is an attribute among mammals that glues family ties, keeps tribes together and prompts a person to risk his life for another. Emotions are communication tools that have helped humans build great cities and great cultures. The little girl in the picture can be considered a communication tool to arouse emotions in another person. Whether it was a propaganda trick by the journalist to enlist help and monetary contribution to that country, I never gave it much thought. It would not have helped me any. It would have come after the water was spilt. A visual stimulus had triggered a very complex physiological and behavioral response process that I could not control at conscious level. At the moment, I feel I must continue pursuing not the tears' motive, but the anger afterwards.

Most people are aware of, or at least, have heard or read about the behavioral trilogy: fight, flight or freeze response. In the case at hand, a visual stimuli was responsible for provoking my initial response of sadness outwardly manifested by tears and wet nostrils. My unconscious must have had connected consonant past painful memories with pain I imagined or interpreted the little girl was experiencing. Repeat, I had no reason to suspect the journalist was using the little girl as a ploy for his own agenda. Sharing pain with thy neighbor is not unique to humans. You can observe elephants getting close to each other making an impregnable wall protecting a wounded calf. Honey bees and wild ants, not the ones you usually make use of for research in laboratories, make self sacrifices for the benefit of the collective unit. We may argue that there is a collective tendency at molecular and cellular level to identify with and rescue a member or a relative of your genus. Of course, we have to consider, in evolutionary terms, when we parted from each other a long time ago. For instance, the mouse, a favorite research tool, its genome sequence is over 90 percent similar to us. However, not many people would shed tears spontaneously over a mouse that was killed while trying to eat his or her expensive cheese. I suppose that you will agree with me that our behavior towards cats and dogs is quite different. Both pets share our food and shelter. Suppose I proceed to identify and name all the neural dynamics that were involved and necessary to produce the corresponding behavioral and physiological response on my part; definitely it would fall short of explaining my experience. The beauty of this argument, if you allow me to call it an argument, is that it provokes me; it motivates me to continue to explore and search for my identity. It leads me to identify and connect with my old self. This old self is my emotional self. This old emotional self must have its place of residence in the limbic system. For the sake of continuing the argument, the limbic system was developed after the primitive old brain stem, but before or simultaneously with regions and cluster of cells in the neo-cortex. The pre-frontal cortex is considered by many neuroscientists as part of the limbic system. A very primitive sub-cortical cluster of brain cells, the amygdala, master of learned fear and door to our emotions, is in constant alert status 24/7. If you add handling chronic stress to it, cells will become dysfunctional or die. There is an amygdala in each temporal lobe. You are a very complex organism.

Why should I go in pursuit of anger if I have not yet found out what triggered my tears to anyone's satisfaction? Well, anger is a very powerful emotion not to try to go to the root of its provocation. We human seem to get angry, often at the slightest provocation. It can get us in serious trouble. I would like to equate anger with poison. It not only clouds our ability to think and act rationally and intelligently, but may lead to serious physical and emotional problems besides pouring vinegar to our social life. If you are an angry salesman, your family may go hungry and the mortgage payment might be in jeopardy. In my profession, I cannot afford to be angry. It is imperative that I pursue the trigger that provoked me get angry at myself for going to the next page in the magazine and try to dismiss my tears as just simple nonsense. Furthermore, tears, sadness and anger can be construed as symptoms of a depressed person. You may think that I am over-extending a seemingly trivial episode with a brain disease. Well, I am going to share with you that a person suffering from a major depression may attempt to kill herself and succeed. She will kill herself when she is experiencing enough strength and anger towards self. She or he might fool the psychotherapist verbalizing she is doing better while in reality she is planning suicide. Now, do not get me wrong, please. I am not equating my experience of anger and tears with a major depression. The thing, if it is a thing, that provoked anger on me must have been a sense of guilt. See, anger, as I said earlier, has its place of residence deep in my brain. It involves, among other group of brain cells, the hypothalamus that will provoke a cascade of hormones mobilizing the entire central nervous system. Do not underestimate the old brain stem with its locus coeruleus and the Ralph nucleus roles in this brain orchestra. The Ralph nucleus located in the upper posterior section of the Pons right above the medulla is a rich nucleus of the neurotransmitter serotonin. A deficit of serotonin is related to depression. Pons comes from the Latin noun, pons which it means a bridge. The locus coeruleus is located near the Ralph nucleus and is a rich nucleus of nor-epinephrine cells. In general, this is an excitatory chemical of this part of the brain stem. As you may have already guessed it, both of these two primitive brain stem group of cells have many connections with the limbic system including the amygdala and hippocampus. The medulla is located, or we can say that the medulla is a prolongation of the spinal cord. These are very primitive group of cells that have existed for a long, long time. The big and carnivorous Tyranosauruses must have a strong spinal cord to be able to move fast and deadly with strong jaws and powerful feet. If I am not mistaken, current crocodiles and alligators with their armored skin, especially, the Nile River crocodile is a descendent of old times huge lizards.

I am not trying to make it hard for you to understand my anger. I am just trying to show you how old are these primitive cells nuclei surviving in our organism. And... quite interesting, in case you did not see it, those two cell nuclei seem to compensate each other. Lack of serotonin from the Ralph nucleus make you feel down, energy-less, and in no mood of enjoying your life and your body. The locus coeruleus, on the other hand, makes you feel excited, and in a challenging mood. I am not suggesting a fighting match between these two cell chemicals, but a life conditioning factor. One may be an attribute for hunters and guerrilla warfare individuals, while serotonin may lead to farming, cooking and home like chores. As far as we know, there is no scientific datum to support this hypothesis. We want to add that besides the biology of depression, there are environmental causal factors that are equally important in the etiology of depression. There are well documented professional reports of infant and childhood motherly love neglect and deprivation conducive to depression symptoms and disease. There is abundant documentation by teachers and psychologist that neglected infants and children do poorly in academic work. They experience sadness, and even physical growth seems slow down. We are not trying to annoy anyone, but if we are seriously looking for a biological trigger for anger, we must include, among others, gene mutations. We also should include epigenetic and environment, diet, illness and harmful habits like alcohol consumption, smoking and illicit drug abuse. A poor diet is mostly associated with people in underdeveloped countries. However, the fast food fad in some rich and developed countries is also a dietary serious problem.

Anger can become a serious health problem because it involves the central nervous system and the peripheral system with myriad of hormones preparing the organism for fight-flight action. It means the heart and the entire blood circulatory system will be engaged in extra work. Arteries will be constricted forcing blood to flow faster. The heart will pump blood out faster sending needed blood to the brain. It may cause headaches or even a stroke if you have an elevated level of low density cholesterol. Needless to say, but the heart itself may become a victim of anger. Epinephrine also known as adrenaline is a major hormone that provokes many physiological responses that identifies, meaning distinguishes stress-induced emotions. Anger, anxiety and fear are a trilogy that seems to have a common denominator. We already mentioned the amygdala as a master cell nucleus for fear and anxiety. Fear, anxiety, anger and aggression can be placed together as survival

181

tools. If we consider anger and aggression as twin emotions that have helped us survive in the jungle, we may appreciate its usefulness. During the expression of both emotions, the organism makes use of a lot of energy facilitated by a powerful and very busy protein in our body, adenosine triphosphate, ATP for short. This protein is engaged in many molecular actions- reactions even going against the normal flow and behavior of other molecules to achieve a specific objective. ATP consists of one molecule of adenosine to which three phosphate groups are attached to it. It does not disintegrate easily while performing its catalyist role. Its high-energy phosphate bonds seem to be unique to this hard working protein. There seems to be hardly any job done in our body that ATP does not donate energy in molecular and cellular functions. ATP releases energy while engaged in its job, therefore, it must replenish itself to resume, meaning going back to its expected role of donating energy. Well, we have sugar in its many different forms, like many other nutrients, is an excellent source of energy that must be transported into cells of all organisms to carry out action. The brain is an organ of large amount of sugar consumption. Our organism needs a permanent supply of energy for its multiple metabolic processes. Our star, the sun, is there for us to make use of it as needed. Our trees and domestic plants like potatoes, bananas, beans, and sugar cane itself are great source of energy for our organism. Also, hydrogen, oxygen, nitrogen and carbon are elements existing in great abundance and available to ATP. When you are running uphill your inseparable friend, ATP, is giving you a hand to finish the job. During its catabolic process, meaning, when ATP is donating- releasing- energy, that energy may be used to do a job in a cell. When ATP, our never tiring and generous energy donating great molecule, is helping other molecules do their job, we may call this process anabolism. While catabolism releases energy, anabolism requires energy. When you kick a ball (anabolism), it goes from a state of inertia into action. When the ball reaches the top and goes downhill (anabolism), it is releasing energy; but when the ball hits your cold glass of soda and spills it, you have another cycle going on. Our organism is a complex and huge factory of trillions of chemicals reactions taking place every second of our life. Atoms share and donate electrons; there are many chemical bonds taking place in molecules and cells in our body making use of the good services of ATP. Learn to love your ATP energy donating macro-molecule. I know you are an intelligent person and may have already thought of any connection existing between ATP and depression. Anger, for one, to be expressed, it needs the mobilization of ATP energy in its multiple forms. Resting ions and molecules need a kick to start rolling and get hormones travel at high speed. We will consider your observation and explore it as soon as possible.

The above named trilogy of fear, anger and aggression has helped us overcome many challenges from the jungle. However, as we evolved into complex societies, anger and aggression had to be tamed and re-channeled to serve us. But when needed, we have to be courageous and aggressive enough to enter into combat with an enemy. Aggression-anger seems to be intrinsic to all animal behavior. We have learned to suppress anger and refrain from overly aggressiveness to form culture and society to share each other and live together. Parental love and care as well as religion have contributed to mold our behavior in a fairly acceptable fashion. A problem may arise when we repress or deny our anger or aggression. We may not be aware that we are denying our anger or aggressiveness. It may have its origin at home when mommy told us to be a nice boy or a nice girl and do not tear apart in anger the cheap toy uncle Joe brought home for a birthday present. Looking for the root of the repressed anger or guilt is not an easy task. After all, it has served us very well for many years. A worldwide goal for the near future is the sequencing of thousands of human genomes to identify many gene variants that will help us identify disease predisposition. Unrestrained aggression- anger has been man's most destructive weapon. It may be considered a disease in need of restraining tools. Our DNA revolutionary technology is fast advancing giving us the answer or answers to questions we are working with now. Gene variations not only secondary to changed base pairs, but also, to histones' modifications through SNPS will surface up as we unravel the genome multiple components segments and their functions.

Michael, a psychotherapist with many years of experience read the first part of this essay and he suggested that modern DNA technology should develop appropriate tools to harness and modify the cluster or clusters of brain cells responsible for aggression. He maintains that those brain cells have lost their usefulness in modern day society. He justifies his argument by pointing out the brain's ultimate weapons-atomic bombs- to defend itself from outside threats. This is the ultimate tool aimed at self destruction. It will not only destroy human beings, but all forms of life on planet Earth. I could not, but agreed with him adding that denial and greed are road blocks on our road on our peaceful journey. You and I are aware that at present time, there are many health hazards that need to be addressed, but the immense majority of people choose to ignore it. Most people are aware that nicotine is a very strong causal and damaging agent on lung disease, but they continue to smoke despite everyday warning. Illegal drug abuse is another health threat among its users not only in the poor echelon of our society, but among the rich, powerful and famous individual in our society. The most common argument by those individuals is that a knife can be used for defense as well as offense, is a very poor analogy. In the use of atomic weapons, almost four billion years of life development on planet Earth would be lost. Men have spent most of his time on this planet fighting each other. Learning to look after and care for his brother fellowmen is a very recent human attribute, perhaps coming out of sharing towns and cities. We are not out of danger yet: slavery, discrimination, hunger, diseases and abuse are rampant in many places on our beautiful and dear planet Earth. Michael's philosophical approach to our problems is sweet music to my ears, but I cannot afford to deviate from my original goal. I must return to the here and now issue: anger, guilt and tears.

Anger is a bug in my brain that, although very useful during my long residence in the jungle, and later on, while establishing towns and building fortresses for my protection, is now a liability. The worse part of it is when anger is turned inward, against my-self. This problem seems to be aggravated when anger is repressed or denied. Our organism defense system is aroused and hormones, otherwise unnecessary, overflow our circulatory system creating an unnecessary hyperactive state of alertness harmful to many vital organs of our body. Glucocorticoids-hormones from the adrenal cortex- are released by bundles in response to a stressful experience. Stress takes over when it needed not to be. Once again, our ability to think, act and behave rationally and intelligently is seriously compromised. "In the hippocampus, a brain region densely populated with receptors for stress hormones, stress and glucocorticoids strongly inhibit adult neuro-genesis... Decreased neuro-genesis has been implicated in the pathogenesis of anxiety and depression..." (Nature, 8/25/11). This same hormone that threatens neuro-genesis may also save the life of an asthmatic person. These hormones are also potent anti-inflammatory medical tools that save life in all parts of the world. An excessive release of this hormone may provoke heart and brain strokes. It goes without mentioning it that the amygdala and prefrontal cortex are also in jeopardy when there is an overflow of this hormone. It just occurred to me now that knowing all the negative ramifications of an overflow of this hormone may make me more susceptible to the release of this chemical in my body. However, it sounds as if I am saying that I should not explore any further assets and liabilities of glucocorticoids. These hormones have assisted me for many years for me to abandon them now. In fact, I am inhaling a simple version of this hormone to control my asthma. Anger may provoke an asthma attack, but I did not experience any problem breathing normally the evening of the little girl chained to a tree trunk. An asthma attack may be provoked by multiple causal factors or triggers including medication you may be taking for another bodily malady. We need to be taught when we are small children that a poisonous snake, an intruding wasp or crazy honey bee sting while we were resting under a mango tree may be lethal. I strongly dislike snakes, and I am not a friend of the other two flying creatures. I have been stung by both flying enemies of mine. If the tree trunk the girl was chained to, had any connection with a mango tree trunk while I was stung by a wasp, I have no idea. A Freudian analyst may insist on exploring this possibility, but I informed you already that I want to go on with a biological basis for my anger following guilt and tears.

I feel I must clarify for my own peace of mind that major depression is considered by contemporary psychiatry and psychology as a brain disease with a biological basis. During my early years of training, psychodynamics had the upper hand and Freudian theory considered major depression a disorder of the mind. The mind could not be the object of medication while psychoanalysis was the preferred treatment modality. Patients used to spend many years on the couch with little progress, if any. It was an intellectual exercise with no personal affective interaction between patient and psychoanalyst. I recall a lady, let's call her Mary, not her real name, who spent years on a couch.

MARY, FROM A SOFT COUCH TO A HARD CHAIR

Mary was a Caucasian 37 year old single female working as a clerk at an airlines office. Her only sibling, a boy two years her junior died when he was seven years old. The cause of his death was leukemia. Mary lived with her parents who took care of their home expenses. Her maternal uncle was the manager of the airline office. She had spent almost six years under traditional psychoanalysis. Her high school years were described as normal, except that she was not actively socializing with the rest of classmates. She had fairly good grades and decided to attend a local community college. Mary wanted to be a school teacher. During her first year in College she did fairly well, but during the first semester of her second year she began to slip down on her work. Mary had lost interest in her education, became easily upset without provocation, had crying spells, experienced sleeping problems, was easily tired and, above all, lost interest in all social life. According to the chart handed over to me by the chief psychiatrist, Mary had had suicidal ideation and vague suicidal plans, but was never hospitalized. She had been raised in a religiously oriented home environment. She confessed her suicidal ideation to the priest. The priest told her that it is a major sin against God to consider self destruction. God had given her life and she must never go against G-d's plans. Mary had gone half way on a bridge connecting Brooklyn and Manhattan and checked out a spot she could jump and kill herself. She said she did not do it because the priest warning on her. She felt she should not disobey G-d. During another occasion, she thought her mother's sleeping pills could be used to finish her unbearable pain and sufferings.

Mary was referred to me by the chief psychiatrist in the hospital I used to work for. He was a European born Freudian trained professional, but with an opened mind. We used to have frank and professional arguments and reservations on orthodox psychoanalytic theory. I was a supervisor carrying almost a regular caseload. I took it upon myself to carry patients regardless of my position at any setting during my professional career. The chief psychiatrist was my immediate supervisor. His office door was always opened to me. The day he called me into his office to assign Mary to me, he said: "It is all yours; she has had enough of the couch". He meant traditional psychoanalytic theory and treatment. My office was unadorned, no pictures to avoid distraction of any kind, except a calendar hanging from a wall. Incidentally, the calendar was showing in its monthly page a yellowish pineapple. It used to remind me of my adolescent hood back home. When pineapples are ripped and ready to be picked up and eaten on the spot, they release a perfume like smell that provokes you to hallucinate a heavenly paradise. Back in my office, the chair I was sitting on had a cushion making my seven hours day work a little more comfortable. The chair Mary was sitting on did not have a cushion; it was purposely arranged that way by me. Each therapeutic session lasted between forty to forty five minutes. When I called Mary in to my office and pointed to her seat, she kind of hesitated and looked around either looking for a couch or a softer seat. She briefly looked at me, and looking at the floor, said: "Are you going to analyze me, are you not?" Mary was neatly dressed; her facial make up was lightly visible. Mary looked again at me with a half forced smile, as if she was waiting for my answer to her opening question. I told her I had read her past treatment history, and that if any analysis had to be done, it had to done by her. I added that any analysis done in my office had to be something that was bothering her at the present time. I further advised Mary that my concern during our 45 minutes session was her feelings as she was experiencing it at the moment. Without any other preparation, I asked how she was feeling sitting on her chair. She seemed to be taken by surprise at my question and apologetically said she was alright. She proceeded to place her relatively long skirt under her legs without looking at me. I said to her: Can you look at me? Immediately after I added: What I mean is how do you feel the chair you are sitting on is pressing against your body? Again, she avoided or skipped getting into feelings and verbalized she attends church services regularly and the pews are not softer than the chair she was sitting on. Of course, Mary was not used to address her own feelings in a way I was asking her to do. She was used to verbalize whatever came to her mind; it had been a mental trip coming from the frontal cortex and left side of her brain. The limbic system with all its emotions was left aside awaiting to be called in if necessary. It was a fragmented approach to repair a brain dysfunctional system. Mary and I had not yet established a trusting working relationship for her to deal with her own self, the feelings that provoked the symptoms for her to come to our clinic for help. A trusting working relationship between a person and the therapist may take several sessions. There are several conditions that may determine how that relationship might go. Resistance to therapy may come in different shades and shapes. It is up to the therapist to lead and point out how to come out of a self destructive behavior a patient may not be aware of. The therapist must temporarily donate his/ her energy so the patient can crawl out of a very slippery emotional hole. I explained Mary that the association she made between the church's pews and the chair she was sitting on in my office was an intelligent one. However, I explained to Mary that the chair she was sitting on is meant for her to feel her body as it relates to her emotions and feelings. In an effort to make the session a little smoother, I looked at Mary and asked her to rest her arms on the chair arms. Rather nervously, Mary looked at me and said she was not used to look straight at a therapist eyes. However, she said her heart was palpitating faster than ever before. I said I was happy to hear that from her. I wondered if she felt her blood traveling all over her body. She paused for a little while, but did not look at me. Afterward, Mary said she felt blood going to her head. Immediately after, she turned her hands up to show me she was sweating. Without saying a word I conveyed to her I was supportive of her. She took a deep breath, and with what appeared to be a spontaneous smile, she said: "Thank you." This first session was over and purposely I said nothing else. I got up from my chair and opened the door for her to leave my office. When Mary had taken the last step out of my office, she turned around and said to me: "You did not tell me when I have to come back to see you again." Mary had gotten my message and wanted to further explore the emotions and feelings that had made her feel ill. She had felt her forearm against the chair, as well as blood flowing in her body. Equally beautiful for her was looking at a therapist in front of her, sharing her fears, her emotions. There was nothing for me to interpret for her. She must have had more than enough interpretations during her prior treatment. What I wanted her to take home from our first encounter was that as an emotional human being, if she allowed me, we could join our energy and deal with dreadful memories that made her come to us for help. She did not have to talk about it over and over again super-activating synapses and forming long term memories. Memories that are already toxic need to be deleted, meaning, deactivated. Building new ones was my goal for Mary. During the following sessions, Mary felt comfortably dealing with feelings as she experienced them during the therapy hour. This therapist was there to assist her get on track whenever an intruding thought provoked her to wander away from herself. Besides, this therapist intervened whenever an emotional overloaded memory demanded extra energy to overcome a severe traumatic episode. Otherwise, Mary had to rely upon her own strength and resources her organism had in storage for her when needed. Mary was able to deal with her feelings towards her parents, particularly, her mother. A religious lady, she verbalized she followed her faith teachings faithfully. Mary also was very respectful of her parents expectations on her. However, during our therapy sessions she verbalized and acted upon feelings of anger towards her mother she had not shared with the priest during confession. During our sessions, she had gained enough confidence in herself to challenge her mother as selfish and ignorant person. Later on, Mary changed it to an over-protective mother who did not know any better at the time. On several occasions, Mary visualized her mother taking part in our session and placed her mother sitting in front of her on top of a toy box I had in my office. Her feelings towards her mother fluctuated between a loving mother and a strict ignorant person. She addressed her mother freely and without hesitation of any sort. In about five month into therapy, Mary asked me if she could come to see me twice a week. Without thinking about it, I told her it was nice to hear she wanted to come for therapy more often. I added that I had to consult it with my supervisor. In reality, I could not make room for her even if I wanted to. Anyhow, I took it over to the chief psychiatrist who pulled out a list of clients waiting to be called in for therapy. He said: "Take as many as you want." Then, in a joking mood, he added: "You are the only therapist asking me for more patients." After we had a good laugh at each other, he continued by adding, I need you supervising all personnel in this clinic. I am busy writing proposals for this one beside our satellite clinic located on Rochester Street. I told him the name of the patient I was referring to. I told him I could accommodate her in my English speaking group. I had another Spanish speaking group. He said to me: "Do as you please, but do not forget what I said about your main job here." I informed Mary about our decision and she looked very happy. In fact, she had already talked with two ladies from the group who welcomed her decision. Mary was coming out her self- built shell and beginning to enjoy the fruits socialization had waiting for her. Later on I learned that some group members used to go out together to Museums and Broadway shows. Before Mary joined the group I explained to her the group existed as if it had a personality of its own. During each group session she could not refer to herself as an individual unit existing outside the group. She had to become and fuse with the group forming a larger collective unit.

During one of our individual sessions, Mary volunteered to work on a dream that was provoking anxiety on her. The dream turned out to be related to her nephew's attempt to play with her genitalia. Her nephew had been invited by her mother to spend the week-ends at home. Mary was aware of my toy box, content and use during each therapy session. While working on the dream, meaning her nephew, Mary went straight to the toy box, pulled out a dummy the size of her nephew and placed it on her seat. She went back to the toy box and pulled out a rubber baseball bat and began hitting the dummy calling his nephew by his name. Fatigued and looking tired, Mary looked at me for approval. I nodded my head with an unspoken yes, go ahead. She pulled the dummy out her chair and threw it on the floor adding in a loud voice: "You do not deserve to be sitting on my chair; get the hell out my head and my life." She celebrated the session saying: "I have wanted to do this for a long time. I am glad I could do it today, thank you." Mary had shed many tears as a child and as an adult. She was also loaded with guilt feelings that had driven her to consider terminating her life. She needed more than talking therapy. She needed to take action against toxic memories that had invaded her brain. Mary continued to attend individual and group therapy until I was promoted to another job in another clinic seven months later. About a year later, my daughter and I went to see a ballet play in Central Manhattan, not far from Broadway. We were on the second floor looking down the nearby plaza. We were early for the show. To my surprise, an elegant, beautiful and well dressed lady came running towards me. Without saying a word, she hugged and kissed me on both cheeks. Right behind her was a handsome and a sharply dressed man about 40 years old. It took me a little while to come to my senses after such a display of affection from a lady who had wanted to die a few months before. She turned around and said: "Tom, this is the therapist I have talked to you about. I want to introduce you to him." While I was shaking Tom's hand, Mary said to me, "Tom is my husband." Not all my cases had a happy ending like Mary had.

Although not a single person undergoing therapy with me ever committed suicide, some required emergency hospitalization. However, I recall a patient of mine, Tito, (not his real name) who came from the same town that I was born. He had been wrongly diagnosed as schizophrenic. His command of the English language was limited, and at the time, Spanish speaking psychiatrists and psychotherapists were few. In our clinic we diagnosed him as bipolar disorder, manic. He was placed on appropriate medication and showed remarkable behavioral improvement. However, improvement for him meant that he was cured and stopped taking his medication. During a visit to my parents, his mother, a friend of my mother, pleaded me to help her son. I promised her I would do everything possible to help him. He had no relatives living in New York City and his friends were very few. By the grace of luck, my mother had given his mother my home phone number. It did not take long for Tito to get my home telephone number. During his manic stage he used to go traveling across the U.S.A. I recall him calling me from Philadelphia, Miami and San Antonio, Texas. His call from Philadelphia was easily solved. I lived four years in Philadelphia. In Miami I had relatives and friends living there who were friends of Tito from childhood. However, they preferred Tito to stay away from their home and families. I was able to arrange a bus ticket for Tito to get to New York City. The funniest call from Tito came from San Antonio, Texas. I had made a visit to that city a couple of years earlier and remembered some important landmarks. He called me saying he was very tired, sleepy, hungry and without a penny on his pocket. When I asked him where he was calling from, he said, Texas. I said, thank you, Tito, but Texas is the second largest State of the Union. When I asked him who was he with, he handed over the telephone to a Mexican lady. She told me it was San Antonio. At the time, there was an agency, Travelers Aid. I knew the location of that agency and directed Tito there. I contacted the agency and they accepted to provide Tito with a one way ticket to New York City after I promised to refund them from our clinic. I have had very interesting individuals during my career as a professional psychotherapist. Brain disorders and brain dynamics has been my concern for about forty years. Whenever I had a patient in front of me during a therapy session, I could not help it, but imagine his or her brain looking like a christmas tree with many lights turning on and off in split seconds. The limbic system including the prefrontal cortex followed by the association areas of the brain were my favorite regions to look for during the therapy hour. Occasionally, I used to say to myself: "Neuron, wake up and get this man in action, help him overcome this impasse." I wanted the emotion to be unlocked so that the patient could use all tools at her/his disposal to weaken down the strength of the toxic memory at hand.

HELGA, THE STAR OF CIRCUS A.

We have talked about DNA, RNA, proteins, the genome and epigenetics followed by the little girl chained to a tree trunk, Jeanne, Mary and Tito. You should not have any problems identifying and dealing with any of the above subjects. The molecules DNA, RNA, proteins and genome are so familiar to us that even during breakfast time they come in different letters, shapes and colors. The patients we just referred to, you can come across them any time walking on the street. We are trying to put together all the above in one solid picture, if possible. We are trying to glue together molecules, protein, cells and diseases into a living picture capable of conveying all the problems we confront and overcome during our lifetime. We know it is difficult, but not impossible. If we fail in our attempt, there is someone out there capable of doing for us. We have talked about the brain, brain diseases and disorders including the victim's pain and suffering when the brain becomes dysfunctional. In retrospect, we believe that we have left out a very important component of the brain puzzle, its chemistry, in particular, the neurotransmitter dopamine. This neurotransmitter is implicated in Parkinson disease, Alzheimer, schizophrenia, depression and addiction besides being the "feel good" king of all brain chemicals. You already know that the largest number of dopaminergic cells is located in the substantia nigra in the pons. We also mentioned for you the ventral tegmental area and the nucleus accumbens as main activation center during drug addiction. We did not forget endorphins (endogenous morphine). We will address it briefly if time and space permits us to.

We remember a patient in our clinic, Helga (not her real name) who seems to me is another good illustration of the brain, brain diseases, bio-chemicals, behavior, medication and psychotherapy. Helga's parents were from northern European ancestry who migrated to U.S.A. not long ago. Her father, John, was a truck driver and her mother, Karen, was a waitress at a large and busy restaurant when they met for the first time. Helga and her parents lived next to Karen's mother's home. They had rented a small, but comfortable house near Helga's grandmother. Helga's father, John, had to stay away from home overnight when he had to make deliveries a long distance from home. When Helga was five years old her mother died. Her maternal grandfather had died when Helga was three and a half years old. When Helga became an orphan she was taken care of by her grandmother for a few months. Her grandmother was diagnosed with lungs cancer that soon metastasized. Her father placed Helga on adoption sponsored by a local religious organization. For the next ten years Helga seemed to have made adequate adjustment under the care of her adoptive parents. She maintained sporadic contact with her putative father. However, after Helga reached her 16th birthday, she began to have provocative arguments at home and at school.

Three weeks before her adoptive parents were ready to celebrate her 17th birthday, Helga left home. Her adoptive parents and her father made all attempts possible to find Helga and bring her home. They were told that Helga was seen at a local gas station carrying a small backpack. She had gotten into a car with two boys and another young girl. Unknown to her father and adoptive parents, Helga had joined a small circus that went from town to town with little success. The main feature on the circus was acrobatic stunts like jumping from trapeze to trapeze and walking the tight rope. There were no elephants or smart chimpanzees and talking birds to amuse the children. Helga had become very good at jumping from trapeze to trapeze and enjoyed her job a lot. Helga had become the main attraction with her daring and spectacular jumps. She was so happy being the main attraction of the show that she contacted her father and adoptive parents. Her father offered Helga to open a restaurant in a nearby town for her to run it. However, the offer was not appealing to Helga. She was the star of the circus and cheers made her feel like a princess.

Helga had had a rough time during the death of her mother and grandparents. She had caring and loving adoptive parents, but the emotional scars left on her by the death in her family were mind- bugging. The cheers and long ovations she was getting from the circus crowd was, in a way, compensating for the emotional vacuum that provoked her to run away in the first place. For her, the circus job was more than a simple job; it was an emotional feedback and resource she did not want to miss. Besides jumping from trapeze to trapeze, Helga had developed a set of daring numbers that made the public call for Helga over and over again after each performance. Money was coming in a way no one had dream of before. The circus manager decided to bring another girl into the show. The new girl, Molly, was not a trapeze competition girl for Helga, but Molly's beauty, charisma and magic tricks she pulled out her hat and brassiere became Helga's most hated acts in the circus. She began to blame and accuse the new girl and the manager of plotting against her. It seems that the guardian angel that had protected her was abandoning Helga for the new girl. And like all disasters in life, the worst takes place when you are less prepared to overcome it. Helga wanted to outshine the new girl by taking jumps she had not done before. Each time Helga performed a sensational feat the public stood up cheering her audacious number. Actually, Helga was not competing against Molly. There was nothing going on between the manager and Molly. Helga was fighting her own delusions; she was competing against herself. And…when you fight against yourself and you are not aware of it, your body is the battleground. Even her manager cautioned her against some of her numbers; but it only reinforced her suspiciousness on him. The last day for Helga's performance in the circus was coming soon. It was a Saturday evening after she had an argument with the new girl. There was anger reinforced by more adrenaline running through her blood invading every cell in her body and brain. Dopamine, the feel good neurotransmitter had gone underground and adrenaline was in command. Logic and rational thinking attributes normally assigned to the frontal cortex ran for cover at the sight of an avalanche of adrenaline. The locus coeruleus nucleus in the brain stem was pumping out its neurotransmitter without stop. The old survival brain nuclei had taken over Helga's neo-cortex. Not seriously considering the risk she was taking, Helga's mind was focused on outperforming the new girl. Actually, she was trying to out- perform her own delusional self. No doubt her judgment was seriously compromised at the time when she attempted to do a rollover in the air when jumping from one trapeze to another. Helga missed the trapeze and hit a circus pole fracturing her right arm in two places. She was rushed to the local hospital where surgery was performed on her. After a week of hospitalization, Helga was discharged to her father's home. Her days as a circus star were not only over, but her mind's delusions had begun to take over her brain on a grand scale.

When Helga came to our clinic she had been hospitalized three times for psychiatric problems. Twice she had attempted suicide by taking alcohol and an overdose of her own medication. Her father used to bring alcohol home and locked it in a chest while carrying the key for the chest box in his pant pocket. However, it did not take long for Helga to outsmart her father, especially after he had a couple of drinks and went to sleep. The last hospitalization in a city public hospital was scary and very painful for Helga. She was brought in by the police on charges that she was soliciting money for sexual favors. The psychiatric hospital she was admitted to was awfully overcrowded. During the day, Helga was thrown in a psychiatric ward with another sixteen more persons of different ages and mental disorders. There was no separation of men and women. Most, if not all, psychiatric patients in the hospital were city residents. Some had to be restrained and taken out to be put in isolation because of overt aggressiveness towards other patients, particularly, out of state females. Helga was basically a country girl raised under strict religious observance until she joined the circus. Helga was roughly treated by other psychiatric patients. Besides having a brain disease, she was ill prepared to deal with poor, uneducated, homeless and mentally ill city dwellers. For some patients, a clean bed and three meals a day in the hospital was the best they ever had. She begged her father to help rent a studio apartment for her to get adequate and necessary psychiatric treatment in the city. She had been on several medications including, but not limited to haldoperidol. Helga used to get monthly medication refills without follow up psychotherapy. Several times she just stopped taking her medication and it was up to her father to get her back on medication. Her father was a caring person, but talking and listening to his daughter's inner mind stories was not his best attribute. His hobby was collecting stamps, particularly, from England, Germany and Russia. Both, Helga and her father used herbs, vitamins and minerals as part of and in addition to regular prescribed medications. Carrots cut in small pieces, salvia, saffron, chamomile and Zarza had magic components only Helga and her father knew about it. Both had taken large doses of niacinamide "to keep their mind healthy," While in therapy with us, Helga joined a religious group that kept strict religious rules of behavior besides daily reading and street proselytizing.

When Helga was not quoting the Bible, she was deeply involved in her aristocratic blood lineage. Helga's delusions were very real for her to share with anybody. There were people plotting against her to take away her rightful place in the royal families of Europe. According to her story, she was connected by blood to Alexandra, the Romanov czarina executed by the Bolsheviks in 1917. She was also connected to the English royal family by Queen Elizabeth's husband. Her life history had to be kept in strict secrecy because Bolsheviks and their sympathizers, all fanatic atheists, were always plotting against her. A secret she begged me not to reveal to anyone under any circumstance was that her mother was the daughter of Stacy. Stacy was the secret name of her grandmother, Anastasia outside Russia. That makes her a granddaughter of the late Russian Emperor, Nicholas II. But her delusions did not stop there. By divine right, Helga believed she was blood related to Queen Nofry, short for Nefertiti, wife of Akhenaten. Further hiding her own self identity, Helga claimed her past ancestors came from another constellation called Aldeb, short for Aldebaran. They came ten thousand years ago to teach Earth's humanoids. Helga, with the support of her father, kept a notebook with numbers and geometrical figures she insisted were passed down on to her in great secrecy by her ancestors thousands of years ago. Among her favorite numbers were 5, 3, and 7. In the second place, but not less important were 1, 9 and 6. She mentioned and used number 5 in her book more often than any other number. She tended to reduce names of people and places to five letters. Helga revealed this secret under condition that I will never reveal her identity. According her, her life history can be found in the Apocalypse by John. Using a combination of number 5 assigned to letters like: blood, water, spade, cross, Jesus, souls, sky, stars, light, etc. all in consonance with her five letter name, Helga. For her, number one is the creator; it is alpha and omega. Numbers 6 and 9 are different stages of the same person in its evolutionary trip in the universe. While a person is number 6, he or she is predisposed and provoked by multiple choices leading to and, opposite to each other. Number 9 is the hardest developmental stage to achieve. From stage 9 you can fuse with number one. She had three notebooks full of her stories. She shared with me only a small fraction of her life story because she could read my mind. I was a person she could trust, she said. She began to trust me when she saw an orthodox Christian church cross on the cover of a book on my desk in my office. Helga claimed she had Alexandra's orthodox cross kept at home. During one session, she said I look like an orthodox priest, therefore, she could trust me with her blessed cross. On another occasion Helga made me a member of the group of aliens that had come to teach Earth humanoids.

Our therapeutic sessions with Helga were basically supportive of her efforts to help herself through socialization and reading religious and self help books. We gave her clear, direct and plainly well focused homework for her to do. We asked her to use the local public library dictionaries for the meaning of words she used most frequently to justify her own delusions. We reinforced this behavioral change asking her feedback from members of the religious group she belong to, as well as friends and people on the street. In other words, we placed on her the responsibility for checking out the validity of her false and distorted thoughts and ideas. We do not fight the victim; we help the patient to confront their dysfunctional self. They themselves make appropriate healthy brain connections. Helga was an intelligent lady. She enjoyed reading books other than those provided by her church group. Less fortunate men and women who could not read as well as Helga could, were happy listening to her read for them. In a way, Helga became a star again. Later on, she was registered in a city sponsored vocational training program leading to a secretarial job. Helga's diagnosis was never clear for us. She had periods of time when she was a lucid and well behaved lady engaged in logical and rational conversation. Anyone not familiar with Helga's life history would not suspect of psychosis followed by ideas of grandiosity and persecution. Helga's psychiatric chart appeared with three different diagnostic opinions from different treatment agencies, namely: bipolar disorder, generalized anxiety disorder and schizophrenia, NOS. You may wonder why Helga had three different psychiatric diagnostic impressions. In the absence of a definite biological marker for schizophrenia and bipolar disorder, manic phase, a definite diagnosis may be hard to come. This confusion is more common when patients are taken to a psychiatric emergency room. First, you have psychiatrists in training; secondly, you have to write a diagnosis from scratch. In the majority of cases, patients' past records are not available at the time and place during the diagnostic interview. Most often, these patients are brought to the emergency room by ambulance or police car. Their history input to the diagnostic impression is very limited. You may be wondering why she had been placed on different medications. Medications are prescribed to deal with her diagnostic impression symptoms at the time. She self medicated with over the counter and self help stores. Both, Helga and her father referred themselves to herbal books. Another brain and bodily input was change of ambient and food intake. We must remember the brain's capacity to change and improve-plasticity-taking over functions not done before. We grow new neurons with the potential for new synapses and new memories. We are not entirely hard wired. We spent the best of our time talking about DNA mutations, genome variants and protein genesis. We did not ignore epigenetics and environment shaping the way Helga feels, thinks and lives. She shed tears, became angry and felt guilt, but she did not give up and kept on track trying to overcome a brain disease. Dysfunctional brain cell circuits and pathways influenced by variant causal agents may be brought back to an acceptable equilibrium of brain chemistry and behavior, thus improving the patient's quality of life. We repeat that sour, toxic and chronic stressful life experiences can kill brain cells, especially, in extremely sensitive areas like the hippocampus, amygdale and prefrontal cortex. They are extremely important centers of cells implicated in learning and memory formation. They are risk areas in predisposed individuals. We need to identify with great accuracy, basic determining clinical symptoms in the absence of biological markers, to begin preventive medicine as early as possible. Of course, we need the patient's cooperation. Helga's late erratic behavior at the circus was not easy to stop. She was forcibly driven by chemical impulses impairing her capacity for logical and rational thinking. School education as well as television documentaries and shows aimed at adolescents and parents would be a good start. It is not only the pain suffered by patients and their families, it is also billions of dollars spent attempting to care of them. It is money that could go for research, prevention and scientifically based treatment. We emphasized to our patients the need to prevent relapses and diseases in general. We do not have to have correct answers for all questions. We often feel embarrassed when we cannot come with a satisfactory answer. We spend a lot of energy and time trying to please the public at our emotional expenses. This type of behavior is self destructive leading to illness. Prevention is very helpful during many occasions. Good planning is essential to prevention. You do not need to be a rigid iron rod to plan your day before you encounter stressful conditions. Your brain sets steps for you to accomplish your goals at the same time you make necessary concessions. Before you leave for your job or go on a trip, reaffirm yourself as an intelligent and healthy person capable of making logical and rational decisions. You can choose when to enter a conversation as well as walking away from a poisonous and unhappy individual. You have the power to do so during almost any condition. There are individuals whose brain are wired in a mixed up fashion. You hear college students, professors, legislators, law enforcing officers and many others committing horrifying crimes you would have never suspected of. You have the power, the wisdom to walk away from seriously compromising situations before it harms you. You can make a list of stressful situations you may encounter during the day. You prepare yourself for multiple possible appropriate solutions leaving you in command during any given condition. Remember, by planning ahead of time, you prepare your brain to come out with a solution to a problem that will serve you best. Your brain has solved a myriad of problems you may have considered extremely difficult. You have billions of brain cells making trillions of connection to serve your needs. Do not attach yourself to a limited number of objects at your disposal. When you attach yourself to one objective like in the case of Helga, you are missing all the opportunities in front of you. You can see how individuals attach themselves to cars, jewels, TV sets, drinks and drugs. Greed and power for the sake of power seems to be the greatest sin you commit against yourself. We should not close these good health practices without mentioning once more how the brain gets hooked on pleasure seeking behavior. We mentioned dopamine and endorphines as brain chemicals that may determine whether you get stuck on a repetitive self destructive behavior as opposed to choosing good health and happiness. Once you allow those two chemicals hijack your brain you have given up all chances of becoming a happy person. Your ventral tegmental area and nucleus accumbens will light up like a christmas tree and stay on until it destroys you. Helga and Mary needed to deprogram themselves and leave behind behavioral responses that prevented them from enjoying health and happiness. Toxic and painful memories that are accumulated in your brain from birth have to be modified or removed from your brain. Helga's delusional thinking, a princess with royal blood from Europe's two most powerful dynasties, blocked reality thinking and good judgment. Despite her risk taking job jumping from trapeze to trapeze, her mind was wandering on far- away lands bound to emotional attachments. Even if there was a genetic component in Helga's brain break-down, preventive therapeutic tools could have prevented stressful and painful episodes. As we grow up we make many unnecessary attachments at the expense of our inborn and experimented survival tools we carry through evolution. In today's faddish life style, we over extend our resources becoming embroiled by unnecessary stressful conditions. We go against our brain powerful resources and rely on childish tricks for survival. Like Helga and many others like her, they are in need of skillful professional therapeutic intervention to tear down toxic brain barriers and prepare them for good health and happiness. It may sound hard to believe, but they need to establish new neuronal circuits and networks. Past abnormal behavior seems to be in front of their thoughts and actions preventing them from focusing their attention in a given present situation. Toxic painful memories bring about structural changes in dendrites. Those memories attempt to stay active and monopolize our attention and thoughts. By analogy, toxic painful memories are like parasites that feed and grow attracting healthy thoughts and memories. During our initial therapeutic sessions with Helga, she could hardly stay with me focusing her attention on a specific feeling or thought. Her brain had stored many confusing memories from her alleged royal family entangled with cryptic messages. As far as I was concern, I was talking to an impostor who had taken over Helga's brain. The lady sitting in front of me giving me cryptic messages based on numbers and words gave me the impression that she was not in my office. Her body was sitting in front of me, but mentally and emotionally, Helga was in a far away land. It is not that she did not care about herself and her treatment with me. She was always on time and did not miss a single session with me. We can say that Helga took refuge from daily life challenges by building an imaginary family. To keep up with her thoughts of exaggerated importance, she chose royal families and cryptic messages. By escaping into these mental fantasies, she kept her brain busy. Helga may have felt relieved of every day challenges and needs by escaping into her imagined palatial life. In a way, she was replacing past toxic memories with memories she chose to enjoy. However, there was a big problem to be overcome during working sessions. Escaping to her world of fantasy brought Helga relief from immediate demands around her. She dressed, talked and attempted to behave like royal princesses do. Her behavior at the clinic caught the attention of our staff. They referred to Helga as my aristocratic friend. My immediate task was to bring Helga to feel her body on a rough chair. I could not deprive Helga of her run-away trips to wonder- land. I had to guide her to find and identify her own strength and attributes she was placing on a fantasized princess. She had to find herself beautiful, intelligent, educated, charming, worthy of public acceptance and praise for her talents. Inside my toy box I had a game that looked like a Chinese abacus. It had three horizontal wooden rods holding 9 beads like in a necklace. It could stand by itself on top of my desk. I was trying to get an object she could see and touch during our session. I introduced it as a possible therapeutic tool during our third therapy hour. Helga's curiosity about the toy did not wait. She asked me how the game was played out. I explained to her that I was going to test myself how long we could stay focused on a stated subject or theme. If we were talking about her feelings while she was fitting on an imagined princess' dress and she jumped to another theme, we would mark it by pulling a bead to the opposite side. I placed emphasis on <u>we</u> as opposed to me or she. Gaining Helga's trust and cooperation in therapy was paramount for success. For instance, when she jumped from "I feel sad now" to another topic like "my father's stamp collection", I made a stop hand signal and pulled a bead to another position. Every time we pulled a bead it meant we were not doing anything compatible with psychotherapy. It was during the following fourth session that Helga said: "I have to be careful how I talk; I will stay here as you want me to." From now on I managed to get Helga tell me she had changed our rules to stay on track. It was the beginning of self guided corrective thoughts. Helga was making progress in psychotherapy besides taking her medication as prescribed by our psychiatrist. Helga was beginning to make changes in her behavior. She began to call the Chinese abacus her brain's inner clock. Her mind was transferring attention from her imagined family royalty connections to here and now conversation with me. I must repeat it one more time. My immediate goal was for Helga to stay on her feelings and actually feel herself as a unique person. She must feel the real Helga sitting in front of me. She had to refer to herself in first person singular such as: I feel relax; I am here with you now. I interpreted it as making new synaptic connections. The Chinese abacus, her brain clock and the rules accepted during our sessions began to take her away from imagined overseas trips. When I did not place the abacus on my desk, Helga would go to my toy box, pull it out, show it to me and place in front of us. To my surprise, one day she came to my office carrying a new and bigger Chinese abacus. She said: "This one is for you." Helga's therapy sessions and medication were rediscovering a healthy and happy young lady. Take- home work was part of our treatment contract. She had to get feedback from family members and friends on how she was progressing on our treatment program. Her father was a very supportive person. He used to call her a very intelligent person. During therapy we would go over her compliments. Helga would say aloud, 'I am an intelligent lady.' She had to repeat it several times over. She liked to repeat, 'I have a beautiful ballerina body'. On several occasions, Helga compared me to her father. She felt she did not have to compete against anybody. My demand on her was to stay focused on her body. Emphasis was placed on her body as opposed to her role in an imagined royal house. Helga did not have Molly to compete against or an audience demanding more of her talent. Most important, she did not have to compete against herself. Self competition was replaced by self realization, self identity and self accomplishments. Predisposition to heritable connected brain diseases and problems are not easily overcome. Early identification and professional guidance and intervention are the best working tools we have. It prevents a lot of tear shedding, guilt, anger and pain. We do not need to see children chained to a tree trunk. We do not need to hear that a mother has drowned her children or that an angry adolescent boy shot and killed his parents. Education and prevention are blessing tools we need to apply before a tragedy strikes us.

Please, make frequent visits to pictures we have provided for you to enjoy. Think of it a game, a game with your brain. It is not only a therapeutic tool of mine keeping troubling thoughts away; but an excellent educational tool. Thinking about the damage you would do to a cluster of brain cells if you become hooked to destructive behavior; it would help you stay healthy.

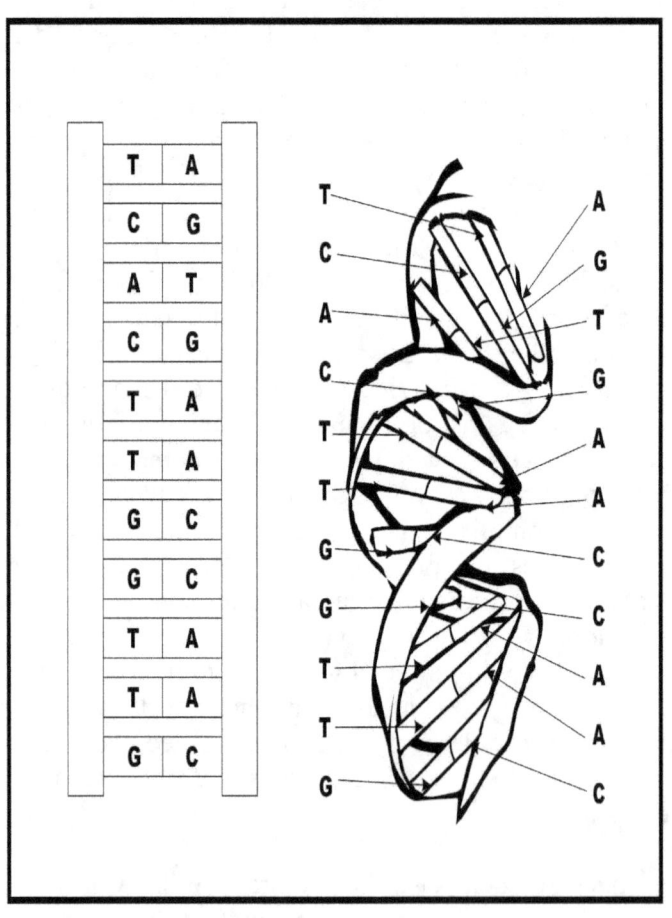

PHOSPHATE-
DEOXYRIBOSE
BACKBONE

5' END

ADENINE

THYMINE

3' END

OH

NH₂

GUANINE

CYTOSINE

OH
3' END

5' END

Play with your brain and learn about yourself. It is fun, you will enjoy it.

Try to identify the neo cortex and the limbic system

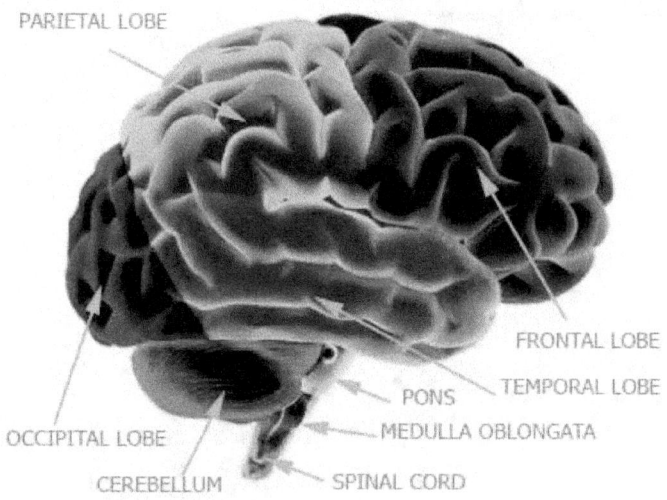

PARIETAL LOBE

FRONTAL LOBE

TEMPORAL LOBE

PONS

MEDULLA OBLONGATA

OCCIPITAL LOBE

CEREBELLUM

SPINAL CORD

Look at this picture; they are neurons in your brain. Have fun and learn. The brain has billions of neurons that communicate among themselves sending impulses across synapses. The message that crosses a synapse is a chemical known as a neurotransmitter. Among neurotransmitters, you have Dopamine, Serotonin, GABA, Glutamate among others. Dopamine is highly implicated in the brain disease of Parkinson, Schizophrenia, Alzheimer and other mental disorders.

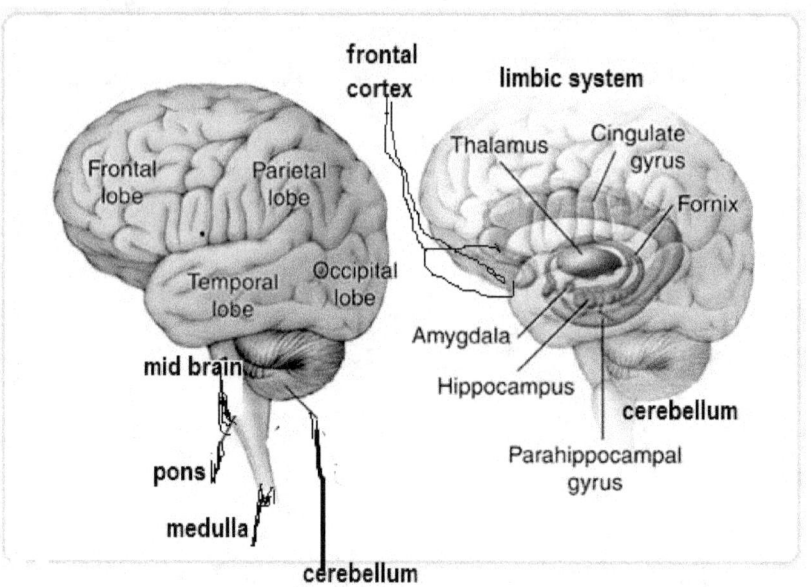

frontal cortex

limbic system

Frontal lobe

Parietal lobe

Temporal lobe

Occipital lobe

mid brain

pons

medulla

cerebellum

Thalamus

Cingulate gyrus

Fornix

Amygdala

Hippocampus

cerebellum

Parahippocampal gyrus

playing with the brain is fun, you
will like it.

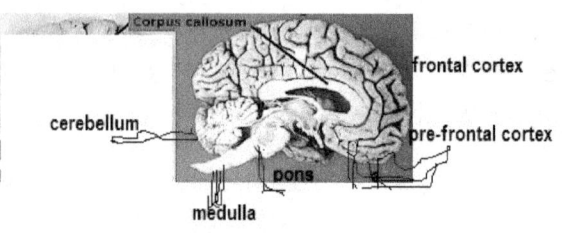

Corpus callosum

cerebellum

frontal cortex

pre-frontal cortex

pons

medulla

playing with the b rain is fun and learning
from other pictures locate the hippocampus, the amygdala
and the thalamus.

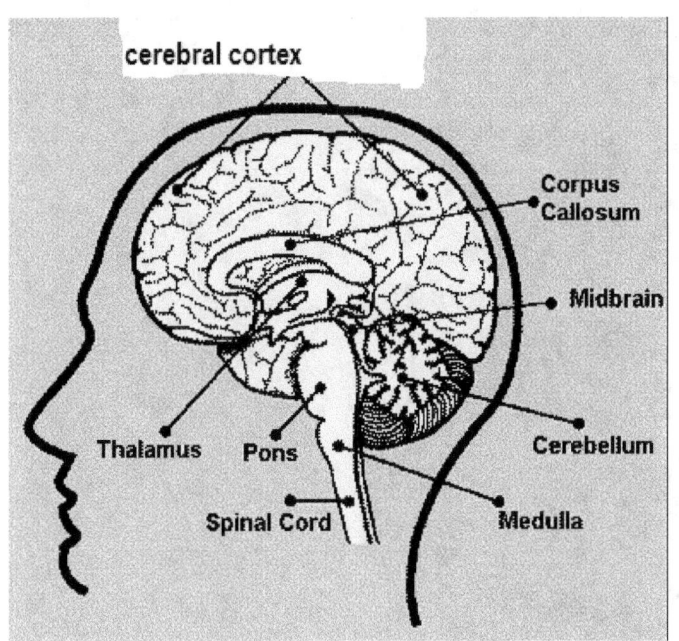

cerebral cortex

Corpus Callosum

Midbrain

Cerebellum

Thalamus   Pons

Spinal Cord   Medulla

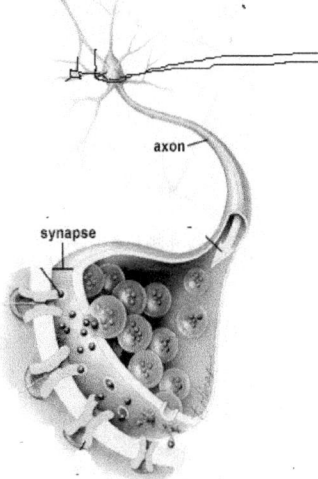

axon

synapse

this is a message coming from another neuron through an axon and received by dendrites by the receiving neuron. notice the structural difference between a dendrite and an axon. among the neurotransmitters we have mentioned several times are: dopamine, serotonin, glutamate, gaba, acetycholine,etc. dopamine and serotonin are implicated in depression. a deficit of dopamine is the main causal agent for parkinson disease. dopamine is also implicated in schizophrenia. not showing in this picture, but the axon is wrapped around by myelin; it helps accelerate electrical speed throughout the   axon

playing with the brain and its billions of neurons is fun; they are all yours to enjoy.

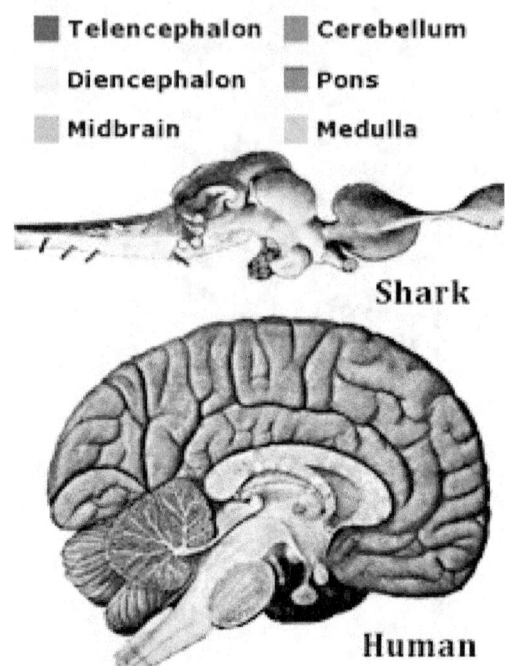

Telencephalon • Cerebellum

Diencephalon • Pons

Midbrain • Medulla

Shark

Human

www.ingramcontent.com/pod-product-compliance
Lightning Source LLC
Chambersburg PA
CBHW060249290526

45789CB00001B/265